FIRE BUBBLES
AND EXPLODING TOOTHPASTE

MORE **UNFORGETTABLE EXPERIMENTS**
THAT MAKE SCIENCE FUN

STEVE SPANGLER

GREENLEAF
BOOK GROUP PRESS

By their very nature, science experiments fizz, bubble, pop, smoke, erupt, move, change temperature and sometimes produce unexpected results. That's why science is fun, and that's why you need to follow the necessary safety precautions when doing any science activity.

The material in this book is provided for informational, educational, noncommercial, and personal purposes only and does not necessarily constitute a recommendation or endorsement of any company or product.

While the science experiments and demonstrations in this book are generally considered safe and a low hazard, please use care when performing any science experiment. Adult supervision of kids is always recommended. If you're an adult who acts like a kid, find a more responsible adult to supervise you. Common sense and care are essential to the conduct of any and all activities, whether described in this book or otherwise. Without limitation, no one should ever look directly at the sun, attempt to build a time machine, or eat yellow snow.

The author and publisher expressly disclaim all liability for any occurrence, including, but not limited to, damage, injury or death, which might arise from the use or misuse of any project or experiment in this book. Wow... this is serious.

Published by Greenleaf Book Group Press
Austin, Texas
www.gbgpress.com

Distributed by Greenleaf Book Group LLC

For ordering information or special discounts for bulk purchases, please contact Greenleaf Book Group LLC at PO Box 91869, Austin, TX 78709, 512.891.6100.

Design and composition by Greenleaf Book Group LLC
Cover design by Greenleaf Book Group LLC
Edited by Debbie Leibold
Photography by Shawn Campbell and Bradley Mayhew

Cataloging-in-Publication data
(Prepared by The Donohue Group, Inc.)
Spangler, Steve.
   Fire bubbles and exploding toothpaste : more unforgettable experiments that make science fun / Steve Spangler.—1st ed.
   p. : col. ill ; cm.
   ISBN: 978-1-60832-189-6
   1. Scientific recreations—Handbooks, manuals, etc. 2. Science—Experiments—Handbooks, manuals, etc. I. Title.
Q164 .S626 2011
507.8   2011935694
ISBN 13: 978-1-60832-189-6

Part of the Tree Neutral® program, which offsets the number of trees consumed in the production and printing of this book by taking proactive steps, such as planting trees in direct proportion to the number of trees used: www.treeneutral.com

Printed in China

14 15 16 17 18 19   10 9 8 7 6 5 4 3

First Edition

CONTENTS

MUST-SEE SCIENCE   95

THE SUPER-SECRET TEACHERS-ONLY SECTION   117

BEYOND THE FIZZ: HOW TO GET KIDS EXCITED ABOUT DOING REAL SCIENCE   151

WHO CAME UP WITH THIS STUFF?   154

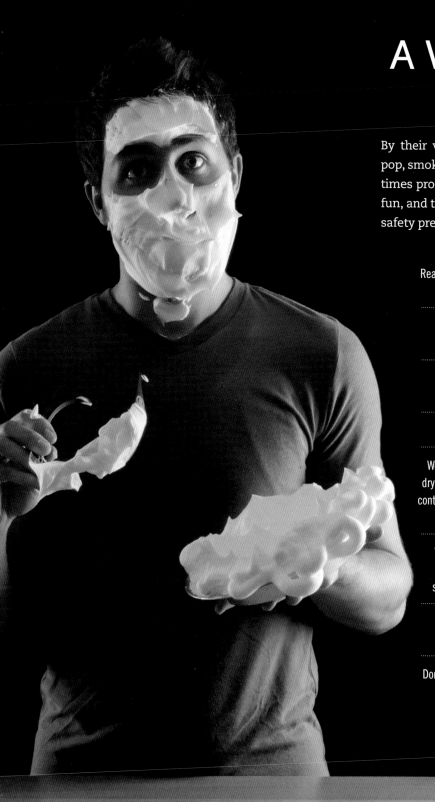

A WORD ABOUT
SAFETY

By their very nature, science experiments fizz, bubble, pop, smoke, erupt, move, change temperature, and sometimes produce unexpected results. That's why science is fun, and that's also why you need to follow the necessary safety precautions when doing any science activity.

Read all the directions before you begin any experiment. If you aren't sure about something, ask!

Take it seriously when the experiments say that they require adult supervision.

Don't put any chemical in or near your mouth, eyes, ears, or nose.

Wear safety glasses.

Wear protective gloves and use tongs when handling dry ice because it will cause severe burns if it comes in contact with your skin. Never put dry ice into your mouth! Never trap dry ice in a container without a vent.

Wash your hands thoroughly with soap and water after handling raw eggs. Some raw eggs contain salmonella bacteria that can make you really sick.

Aim anything that is going to "shoot" or explode away from yourself and others.

Don't eat your science experiments . . . they don't taste good, and eating in a lab is a bad thing to do.

DON'T TRY THIS AT HOME . . .

Let's cut to the chase and be honest . . . science experiments have changed over the years. Okay, maybe the experiments have not changed, but the way they're packaged has. It seems that all of today's science experiments come with a warning that reads, **"Don't try this at home!"** This is especially true when someone breaks out the vinegar and baking soda or anything else that might fizz, bubble, pop, or get a child excited about learning.

What's the first thought that pops into a young scientist's mind when she hears, "Don't try this at home?" That's right . . . "I must do everything possible to try this at home!" The warning issued by the adult becomes a challenge for every kid who hears it.

There's an alternative to the standard warning. Instead of messing up your *own* home, try messing up your *friend's* home. Now the warning reads as follows:

"Don't try this at home . . . try it at a friend's home!"

Here's the good news. This book is filled with great science activities, demonstrations, and science fair project ideas that are easy to do and are guaranteed to get your creative juices flowing. Don't be fooled by the list of simple materials—vinegar, eggs, plastic bags, salt, soap—required for many of the experiments. Even though they're basic ingredients, the "wow" factor of the activities is huge. At the end of each activity, you'll learn the real science behind all of the "gee-whiz." You'll learn not only the *how* but also the *why*. And then something strange will happen. You'll start to ask your own questions and create your own experiments. Don't be surprised if a little voice in your head starts to ask things like, "What would happen if I changed this or tried that?" Curiosity will get the best of you and you'll find yourself doing the experiment again and again with your own changes and ideas.

My favorite part of each activity in this book is the section titled "Let's Try It!" It's not a suggestion . . . think of it as a command or your marching orders. Round up the supplies, clear the tabletop, put on those safety glasses, and get to work. What happens next is the best part. Out of the clear blue you'll make a new discovery and uncover your own science secret. You'll feel your heart start to speed up and your mind race with new ideas. You've made a discovery, and that's an amazing feeling.

—Steve Spangler

AIR-MAZING

WINDBAG **WONDERS**

Here's the challenge … How many breaths would it take to blow up a 8-foot-long bag? Depending on the size of the person, it may take anywhere from ten to fifty breaths of air. At the end of the challenge, the person is totally out of breath, wondering why she said yes in the first place. Now imagine the look on her face when you are able to inflate the giant bag using only one breath of air. That's right … one breath and you win! This is one of my all-time favorite science demonstrations, and it's guaranteed to make it into your Top Ten list.

WHAT YOU NEED

Windbag (available at www.SteveSpanglerScience.com)
Or
Diaper Genie® bag

NOTE: The "Windbag" is actually a long plastic bag in the shape of a tube. SteveSpanglerScience.com is your source to purchase the brightly colored Windbags pictured throughout the pages of this activity. There is a real-world version of a Windbag at your local department store. Head to the aisle where baby products are sold and look for a diaper disposal system (commonly referred to as a Diaper Genie). The long plastic bags are sold as refills for the diaper disposal system, and they work very well for this demonstration.

LET'S TRY IT!

1. If you're using one of the Diaper Genie bags, cut off a section of the plastic tube material that is roughly 6 to 8 feet long. A shorter section of bag (4 to 5 feet long) is recommended for younger kid-scientists.

2. Tie a knot in one end of the bag. Invite a friend to blow up the bag, keeping track of the number of breaths it takes. Then, squeeze all of the air out of the bag. Explain to your friend that you can blow up the bag in one breath. Chances are, he or she won't believe you, but that's all part of the surprise.

3. Have your friend assist you by holding onto the closed end of the bag. Hold the open end of the bag approximately 10 inches away from your mouth. Make the opening as wide as you can with the index fingers and thumbs of both hands.

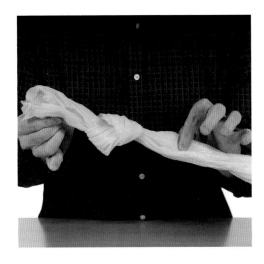

Using only one breath, blow a long, steady stream of air into the bag (just as if you were blowing out candles on a birthday cake). You MUST keep your mouth off of the bag (about 10 inches away from the opening) and keep the opening of the bag as large as possible. As you'll soon see, the secret is actually in the open space between your mouth and the bag.

4. If you do it correctly, you'll see the bag rapidly inflate. The trick is to quickly seal the bag with your hand so that none of the air escapes. Tie a slipknot in the end of the bag or let the air out and try again.

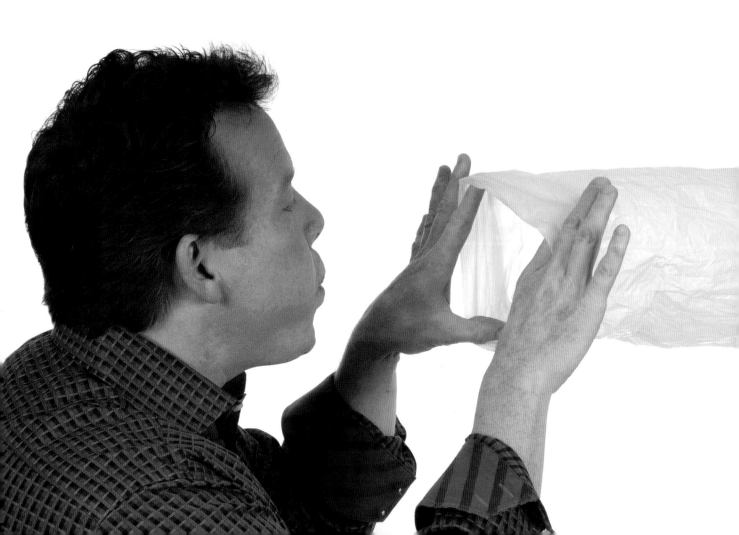

The Ultimate Windbag Challenge

Announce to your audience that they have 5 minutes to work together to build the largest freestanding Windbag structure they possibly can. The structure must be held up only by the Windbags themselves—no one can physically hold up the structure. *Note: It would help to do this activity in a gym, a large ballroom, or outside.*

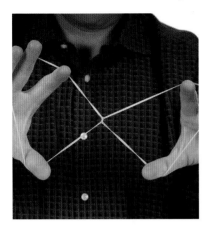

Here's a tip: loop two rubber bands together to form a "figure eight." Now hook two Windbags together by slipping the rubber bands over the tied ends of two inflated Windbags. Use more rubber band "figure eights" to connect multiple Windbags and create all kinds of creative structures. It's a great team-building activity for kids and adults alike.

WHAT'S GOING ON HERE?

Here's the quick and simple answer. The long bag quickly inflates because air from the atmosphere is drawn into the bag along with the stream of air from your lungs.

For you science enthusiasts out there, here's a more technical explanation. In 1738, Daniel Bernoulli concluded that a fast-moving stream of air is surrounded by an area of low **atmospheric pressure**. In fact, the faster the stream of air moves, the more the air pressure drops around the moving air. When you blow into the bag, higher-pressure air in the atmosphere forces its way into the area of low pressure created by the stream of air moving into the bag from your lungs. In other words, air in the atmosphere is drawn into the long bag at the same time that you are blowing into it.

REAL-WORLD APPLICATION

Firefighters use **Bernoulli's Principle** to quickly and efficiently force smoke out of a building. Instead of placing fans up against a doorway or window, a small space is left between the opening of the building and the fan in order to force a greater amount of air through the building. Firefighters call this "positive air flow."

STEVE SPANGLER USES WINDBAGS TO SET **GUINNESS WORLD RECORD**

Steve used Windbags to demonstrate the power of air at the first annual 9News Weather and Science Day at Coors Field in Denver, Colorado, on May 7, 2009. As part of Weather and Science Day, Steve Spangler Science was awarded the Guinness World Record for the Largest Physics Lesson, with 5,401 participants using their own Windbags to perform an independent physics activity. Participants had 2 minutes to inflate their Windbags, and the news helicopter hovering above the stadium captured the colorful scene. Danny Girton Jr., official adjudicator for Guinness World Records, was on hand to verify the record-breaking event and presented Steve Spangler and his team with an official Guinness World Record certificate at the close of the day.

THE QUICK-POUR **SODA** BOTTLE RACE

Race to see who can be the first to empty a soda bottle full of water. With a special twist of the hand, you will be able to empty the water in the soda bottle in just a few seconds.

LET'S TRY IT!

1. Remove the label from the soda bottle so you have a clear view of the inside. Fill the soda bottle almost to the top with water.

2. Without squeezing the sides of the bottle, turn it over and time how long it takes to empty all of the water. Just hold the bottle upside down. You might want to repeat this several times and average the results. Be sure to use the same amount of water for each trial. Now you're collecting data!

3. Keep a table of the trials and call this the Glug-Glug Method.

4. Refill the bottle almost to the top with the same amount of water as you did before. When you turn it over this time, move the bottle in a tight, clockwise or counterclockwise circular motion as the water pours out.

5. Keep moving the bottle like this until you see the formation of what looks like a tornado in the bottle. The water begins to swirl, a vortex forms, and water flows out of the bottle very quickly.

6. Time this method as you did before and call it the Vortex Method. Repeat the test several times and average the results. Which method allows the water to exit the bottle more quickly?

TAKE IT FURTHER

See if you can figure out new methods for getting the water out quickly. Time your trials and record them. Get another bottle and challenge your friends to a race. Until they learn the secret, you will win every time.

For a giant tornado, try filling a Deep Rock brand water bottle to the top with water, swirl it quickly in a clockwise or counterclockwise circular motion, and watch the powerful vortex form. Okay, this method is going to take a very strong person to swirl the water . . . and it's going to make a huge mess. Just take it outdoors and enjoy watching the strong person's shoes get drenched!

If you want to create the tornado over and over without having it drain down the sink, try a popular science toy called the Tornado Tube® (available at www. SteveSpanglerScience.com). The Tornado Tube connects two plastic soda bottles together and allows for the water to move from one bottle to the other as the bottles are tipped. Start with all the liquid in one bottle, quickly tip the bottle upside down, and start the swirling motion. The tornado (it's actually a vortex) will form as the liquid moves into the bottom bottle.

If you want the tornado to be more visible, squirt a few drops (no more) of liquid soap into a bottle of water. Connect the bottle to an empty bottle using the Tornado Tube, shake the bottles to make some suds, and swirl the liquid quickly in a circular motion. Look for the vortex in the middle of the bottle. The drops of soap help make it more visible.

Sure, anyone can color the water by adding a few drops of food coloring. Here's a real challenge: how would you go about coloring just the swirling vortex while keeping the surrounding water colorless? I struggled with this self-imposed challenge for months. My initial thought was to try adding a small amount of oil to the water. Of course, it's next to impossible to add coloring to oil, and the thickness (or **viscosity**) of ordinary vegetable oil destroyed the formation of any kind of vortex.

Then I stumbled upon a kind of colored oil at the local hardware store. It's called lamp oil, and it's used in outdoor lanterns or indoor oil lamps. Best of all, lamp oil comes in an assortment of colors. Purchase your favorite color of lamp oil (the red oil makes a really cool colored vortex) and try adding 2 ounces of the oil to the water in the soda bottle. Use the Tornado Tube to connect the two soda bottles and swirl the water using your now famous vortex-forming, swirl-of-the-bottle technique. When the oil and water swirl together, the less dense oil travels down the vortex first and creates a colored tornado effect. Remember, oil and water do not mix because oil is **hydrophobic** (water-fearing). The two liquids are said to be **immiscible**, which means oil and water cannot be mixed or blended. Since the oil is less dense than the water is, it forms a layer that floats on the surface of the water.

As long as you're adding things to the water, go on a scavenger hunt for a few miniature plastic houses from an old Monopoly game, plastic barnyard animals, glitter, beads, and anything else you can think of. Place the items in an empty bottle and fill the bottle three-fourths full of water. Attach the Tornado Tube and the second bottle. Swirl the liquid to create the vortex and watch what happens to the items you put in the bottle. Where were the items before you swirled and where did they go once the tornado formed? Toto, we're not in Kansas anymore!

WHAT'S GOING ON HERE?

"Auntie Em, Auntie Em, it's a twister!" Well, it's *sort of* a twister. If you've ever seen a dust devil on a windy day or watched the water drain from the bathtub, you've seen a **vortex**. A vortex is a type of motion that causes liquids and gases (both are fluids) to travel in spirals around a centerline. A vortex is created when a rotating liquid falls through an opening. Gravity is the force that pulls the liquid into the hole, and the rotation causes a continuous vortex to develop.

Swirling the water in the bottle while pouring it out causes the formation of a vortex that looks like a tornado in the bottle. The formation of the vortex makes it easier for air to come into the bottle and allows the water to pour out faster. If you look carefully, you will be able to see the hole in the middle of the vortex that allows the air to come up inside the bottle. If you do not swirl the water but just allow it to flow out on its own, then the air and water have to take turns passing through the mouth of the bottle (thus the glug-glug sound).

MYSTERIOUS WATER SUSPENSION

At first glance, you might think that you've seen this science demonstration attempted by a friend. A glass jar is filled with water and covered with an index card. The whole thing is turned upside down and the hand that is supporting the index card is pulled away. The card appears to be stuck to the inverted jar of water. In and of itself, this is a very cool trick, but in this version of the experiment, things get crazy. The inverted jar of water is held over the head of a spectator and the index card is pulled away! The gasps are audible . . . someone screams in anticipation of the water falling from the jar only to drench the poor spectator. To everyone's amazement, the water does not fall. It's suspended in the jar, literally floating above the spectator's head. The card is replaced, the jar is returned to its upright position, and the science magician pours the water back into the pitcher. The only sound you can hear is that of people scratching their heads. Amazing!

WHAT YOU NEED

Mysterious Water Suspension Kit
(available at
www.SteveSpanglerScience.com)
Or
Mason jar (pint size)
with twist-on lid

Plastic screen mesh (This
is the stuff used to make
a screen for a window.)

Scissors

Index cards

Towels to clean up your mess

LET'S TRY IT!

1. Believe it or not, the secret to this science magic trick was in plain view of the audience the entire time. There's simply a piece of mesh screen that is held in place by the lid of the jar. Unlike a normal jar lid, the Mason jar has a lid that comes in two pieces—the center section and an outer ring called the sealing band. You will only be using the outside ring portion of the lid for this science trick.

2. Place the plastic screen mesh over the opening of the jar and twist on the ring portion of the lid. Using scissors, cut around the lid to trim off the edges of the screen. If you want a more professional look, remove the lid before cutting the screen. You'll see that the lid leaves an indentation in the screen material. Use scissors to cut around the indentation. What you're left with is a screen insert that fits perfectly into the top of the sealing band. Place the screen over the opening of the jar and twist on the lid.

3. Your first inclination might be to try to hide the screen from your audience, but the truth of the matter is that no one will see it unless they know to be looking for it. Of course, you'll need to have a little distance between you and your audience, but you can casually show the top of the jar in one hand while picking up the pitcher of water in the other and no one will suspect a thing. Go ahead, try it, and you'll be amazed that you got away with flashing the secret right before their eyes! This is called "misdirection" and it fools the audience.

4. When you're ready to perform the trick, fill the jar with water by simply pouring water through the screen.

5. Cover the opening with the index card. Hold the card in place as you turn the card and the jar upside down. Let go of the card. Surprisingly, the card remains attached to the lid of the upside-down jar. Carefully remove the card from the opening and the water mysteriously stays in the jar!

6. Replace the card, turn the whole thing over, remove the card, and pour out the water . . . while enjoying the sounds of ooohs and ahhhs!

TAKE IT FURTHER

Experiment with different screens, some with fine mesh and some with coarse mesh, to observe how surface tension and air pressure work together to accomplish this feat. Ultimately, it's best to use plastic screen material since it will not rust or discolor the jar. Test out different kinds of plastic mesh from produce bags, for example, to see how the size of the mesh affects the surface tension of the water.

If you want to be really tricky, prepare one jar with the screen and one without. Ask a volunteer to join you on stage and have the volunteer use the jar without the screen. While your jar mysteriously holds the water, the volunteer's jar loses its contents every time. After the laughter subsides and before your volunteer's confusion turns to frustration, reveal the secret . . . but make sure you have a towel close at hand.

WHAT'S GOING ON HERE?

This is truly an amazing science magic trick because several scientific principles come into play to make the water appear to be suspended in the jar. **Atmospheric pressure** (the pressure exerted by the surrounding air) is the force that holds the index card in place. The card stays on the upside-down jar because the pressure of the air molecules pushing up on the card is greater than the weight of the water pushing down.

But how does the water stay in the jar when the card is removed? The answer is **surface tension**. The surface of a liquid behaves as if it has a thin membrane

stretched over it. A force called **cohesion**, which is the attraction of similar molecules to each other, causes this effect. The water stays in the jar even though the card is removed because the molecules of water are joined together (through cohesion) to form a thin membrane between each tiny opening in the screen.

If you tip the jar at all, air will come into the jar and break the seal, causing the water to pour out. Tip the jar sideways and the water falls out of the jar. If you return the jar to its upright position, the air can no longer get into the jar and the rest of the water will stay inside. Now you have a special insight—be careful not to jiggle the jar or touch the screen because you'll break the surface tension and surprise everyone with a gush of water. Okay, maybe that's actually a good idea.

REAL-WORLD APPLICATION

After a rainstorm, you might have noticed that the screens on your windows at home are saturated with water. It's a simple matter of water molecules holding onto the screen (this is called **adhesion**) while holding onto each other and stretching across the tiny openings of the screen mesh (cohesion) to form a thin layer of water. Run your fingers across the screen and what happens? You break the surface tension of the water.

If you've ever gone camping and you had a tent with a screen opening at the top, you might have experienced an accidental rain shower inside your tent. During the night, moisture in the air condenses on the screen, filling each mesh opening with water. Jiggling the tent in the morning knocks the water loose, and you're left with a tent full of unhappy campers. Make sure you try it on someone else's tent!

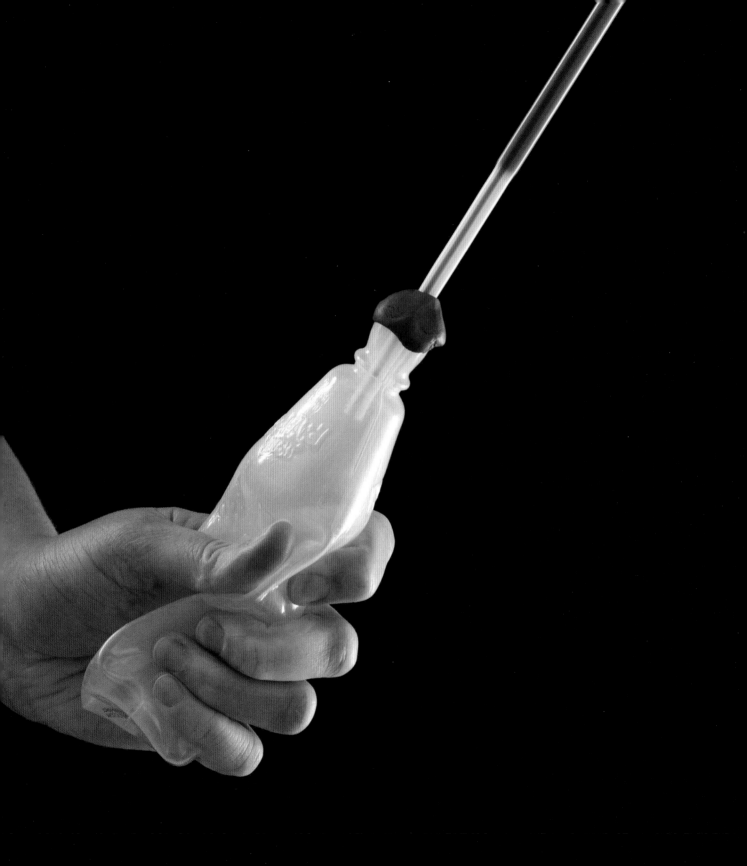

SQUEEZE-BOTTLE
STRAW ROCKETS

It's easy to turn a juice bottle into a rocket launcher. How? Grab a few straws, some modeling clay, and an empty juice bottle to make a launcher that will send the straw rocket soaring across the room. Okay, you'll learn something about Newton's Laws of Motion at the same time.

WHAT YOU NEED

Kool-Aid Bursts juice bottle (flexible plastic bottle)

Modeling clay

2 straws (Find one that's large in diameter and one small. The larger diameter straw must be able to slip over the smaller straw. The large and small straws from a very popular coffee shop chain work great. Follow a star and bring a few bucks.)

Styrofoam packing peanut

LET'S TRY IT!

1. Drink the juice! Enjoy it and then clean and dry the bottle.

2. Push the smaller straw into the opening of the bottle. The straw should fit snugly in the hole at the top of the bottle. Leave as much straw as possible sticking out of the bottle to hold up the larger straw.

3. Use modeling clay to seal any possible leaks between the straw and the hole in the bottle. The clay will also make the straw more stable and less likely to wobble.

4. Push one end of the bigger straw into another piece of modeling clay. This "plug" will seal the end of the straw. Cover the plugged end with something soft, like a Styrofoam packing peanut, to keep the straw rocket from hurting anyone in case they get hit ("accidentally," of course).

5. It's time to launch. Place the larger straw over the smaller straw. Ready, aim, squeeze! The larger straw launches off the smaller straw and the room erupts in a chorus of ooohs and ahhhs!

WHAT'S GOING ON HERE?

While you're having fun launching straws, you're actually learning about **Newton's Laws of Motion**. According to the **First Law**, an object at rest (the larger straw) wants to stay at rest if it's not moving and to keep moving in a straight line if it is moving. That straw will not move unless some force is applied to it to make

it move. That's where your actions come into the equation by squeezing the juice bottle. **Newton's Third Law** says that for every action there is an equal and opposite reaction. As you squeeze the bottle, air is forced out of the smaller straw and pushes against the clay plug in the larger straw. The resulting force causes the straw to "launch" through the air.

Be careful! Never point the straw rocket at anyone. The goal here is to launch the rocket up in the air (not at someone). Be sure to cover the plugged end of the straw with something soft and round to protect someone from accidentally getting hurt by a sharp edge. Be creative! Once you've mastered the simple straw rocket, challenge your friends to a straw rocket design contest. Add a nose cone, some fins, a few decorations, and don't forget to name your straw rocket.

POP BOTTLE MUSIC

A popular Las Vegas musical act uses tubes, bottles, trash cans, and other common items to make some very cool and distinctive sounds. This just proves that banging on pots and pans can lead to a very successful music career. Try your hand at making your own instruments using just a few household items.

WHAT YOU NEED

8 glass bottles (All of the bottles need to be the same.)

Water

Spoon

Good ears

LET'S TRY IT!

1. Fill one bottle full with water and leave a second bottle empty. Use the back of the spoon to gently clink both bottles. How are the sounds different?

2. Fill a third bottle half full with water. Clink all three bottles. The sound of the half-full bottle is about in the middle of the other two sounds.

3. Blow air across the tops of all three bottles. What do you notice?

TAKE IT FURTHER

By varying the amounts of water in each bottle, it's possible to create a musical scale. That's why this activity calls for eight bottles, one for each note of the musical scale. Try it with clinking the bottles and with blowing over the tops of the bottles. What differences do you notice? If you want to really put on a show, use food coloring to color the water in each bottle differently. Of course, the food coloring does nothing to affect the sound, but it does make it look like you really know what you're doing! The ultimate goal is to play a song . . . and then to get people to drop a few bucks into your hat. See, this book is already making you money.

Invite some friends over and present them with this challenge: in 60 seconds, arrange the bottles in such a way that when they are clinked with the spoon, they play a familiar song. Try "Jingle Bells," "Mary Had a Little Lamb," "Twinkle Twinkle Little Star," or "Beethoven's Fifth Symphony"—the song is up to you. The first person to arrange the bottles correctly and play the song wins.

WHAT'S GOING ON HERE?

The science of sound is all about vibrations. When you hit the bottle with the spoon, the glass vibrates, and it's these vibrations that ultimately make the sound. You discovered that tapping an empty bottle produced a higher-pitched sound than tapping a bottle full of water did. Adding water to the bottle dampens the vibrations created by striking the glass with a spoon. The less water in the bottle, the faster the glass vibrates and the higher the pitch. The more water you add to the bottle, the slower the glass vibrates, creating a lower pitch.

The same bottle that makes a low-pitched sound when you tap it with a spoon makes a high-pitched sound when you blow across the top. The same bottle produces opposite sounds! When you blow into the bottle, you are making the *air* vibrate, not the glass. An empty bottle produces a lower pitch because there's lots of air in the bottle to vibrate. Adding water to the bottle decreases the amount of air space, which means there is less air to vibrate. With less air, the vibrations happen more quickly and produce a higher pitch.

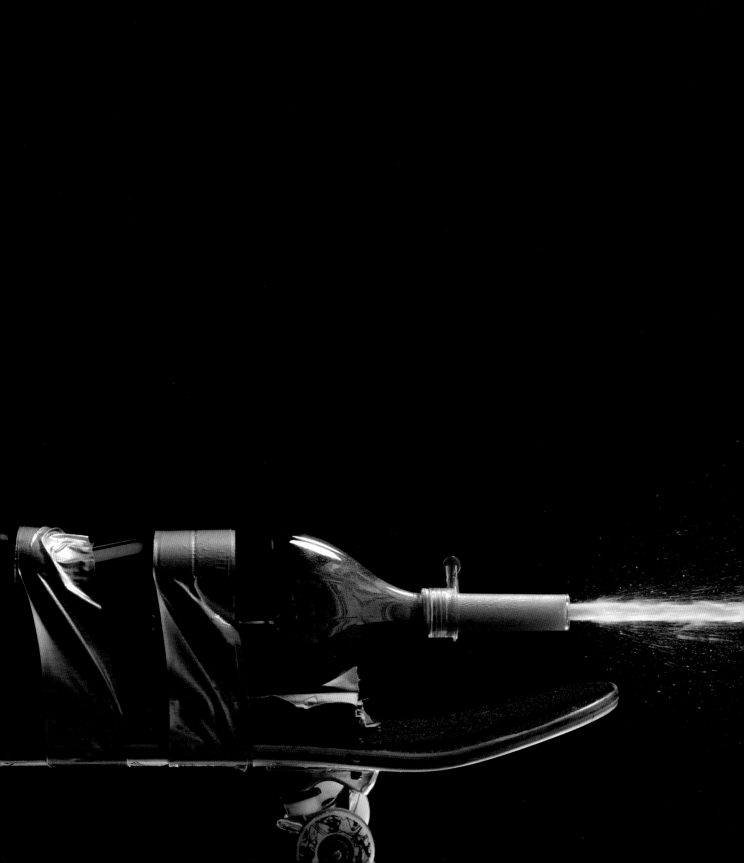

KITCHEN **CHAOS**

18 OZ.
2 CUPS
STATES OF AMERICA
IN GOD WE TRUST
2 HI
1 LO
0 OFF
BLENDER

MONEY IN A BLENDER—
A CASH SMOOTHIE

It's true . . . some money is magnetic. Vending machines will sound an alarm if an ordinary piece of paper is inserted into the slot in place of a real dollar bill. If there's iron in a dollar bill, the only logical question for science enthusiasts is, "How can I get the iron *out* of that dollar bill?"

WHAT YOU NEED

$1 bill (Be sure to borrow it from a friend.)

Kitchen blender

Water

Quart-size ziplock plastic bag

Strong magnet
(a neodymium magnet does an amazing job. Available at www.SteveSpanglerScience.com)

NOTE: Magnets come in all shapes, sizes, and strengths. Ask at your local hardware store for a strong magnet for a science experiment. The strongest magnets in the world that are available to folks like us are called neodymium magnets. Neodymium is a chemical element with the symbol Nd and atomic number 60 *(continued on page 37)*

LET'S TRY IT!

1. You'll need a dollar bill. Now, you could just dig down deep into your own pocket to find a bill, or you could take our advice and borrow the bill from a friend. Hey, why should you have to provide the entertainment and pay for it too? Hold the neodymium magnet near the bottom of the bill. Notice how the bill is attracted to the magnet.

2. Fill the blender one-half full with water (between 3 and 4 cups). Be sure to ask an adult for permission to use the blender.

3. After the dollar bill has been thoroughly examined to verify that it's real, drop the bill into the blender and put on the lid.

4. What's next? Make dollar-bill soup! Grind it, blend it, or liquefy it—just make sure it's torn into thousands of little pieces.

5. After the blender has been grinding away for about a minute, turn it off and pour the contents into the ziplock plastic bag. Seal the bag.

6. Place the neodymium magnet in the palm of your hand and place the bag of money soup on top of the magnet. Place your other hand on top of the bag and rock the slurry back and forth in an effort to draw all of the iron to the magnet. Flip the bag over and look closely at the iron that is attracted to the magnet. You can slowly pull the magnet away from the bag to reveal the iron.

TAKE IT FURTHER

It's easy to suggest repeating the experiment with a $5 or a $10 bill, but don't waste your money. A higher dollar amount doesn't mean a higher iron content.

WHAT'S GOING ON HERE?

It's really very simple. The government uses specially made magnetic inks to print money. This makes it easy for vending machines to "read" the dollar bills, for example, and for banks to determine if money is real or counterfeit. The blender does a great job of tearing up the paper and releasing the magnetic ink into the water. Of course, metallic iron does not dissolve in water; instead, it floats around waiting for a magnet to pull it away from the fibers of the paper.

Here's an interesting question: is it okay to destroy a dollar bill? The destruction (or "mutilation" which has a cool, dangerous ring to it) of paper money by artists, magicians, performers, origami enthusiasts, mad scientists, and everyday folks has been around ever since someone dreamed up the idea of printing bills. So go ahead, blend away! Just don't try to put that cash smoothie back into circulation—or you might find yourself explaining the science of magnets to the Secret Service.

REAL-WORLD APPLICATION

A Meteorite Hit My House!

It's true. Your house has probably been bombarded with hundreds, maybe thousands, of meteorites and you survived the impact. Granted, the meteorites were small in size . . . so small you'd need a microscope to see them, but the house was hit just the same. Chances are, your house has been hit by a few thousand micrometeorites, and you'll be able to find a few if you know the secret place to look.

How to use a super-strong neodymium magnet to find meteorites

The next time it rains, place a bucket under a drain spout in order to collect a good quantity of rain. Get rid of the leaves and other big debris and then sift the remains through a bit of old window screen. What you're after is so small that you'll need a very strong magnet (like a neodymium magnet) to find them. Use this super-strong magnet to determine if any of the remaining particles contain iron. Those particles may be space dust, also known as micrometeorites. And you thought we were kidding!

POP YOUR TOP

What happens when you have a buildup of gas? Wait, on second thought, don't answer that question. The gas in this experiment is nothing more than bubbles of carbon dioxide and the explosion is nothing short of fun.

WHAT YOU NEED

Alka-Seltzer tablets

Film canister with a snap-on lid (Available at www.SteveSpanglerScience.com)

Empty paper towel roll (the cardboard tube) or a similar-sized tube

Duct tape

Construction paper/odds and ends to design a rocket

Water

Paper towels for cleanup— you already know that this one is going to be good!

Watch or timer

Notebook

Safety glasses

LET'S TRY IT!

1. Put on your safety glasses.

2. Divide an Alka-Seltzer tablet into four equal pieces.

3. Fill the film canister one-half full with water.

4. Get ready to time the reaction of Alka-Seltzer and water. Place one of the pieces of Alka-Seltzer tablet in the film canister. What happens?

5. Time the reaction and write down the time. How long does the chemical reaction last? In other words, how long does the liquid keep bubbling? Why do you think the liquid stops bubbling? Empty the liquid from the film canister into the trash can.

6. Repeat the experiment, but this time place the lid on the canister right after you drop in the piece of Alka-Seltzer. Remember to start timing the reaction as soon as you drop the piece of Alka-Seltzer into the water. Oh, by the way, stand back! If you're lucky, the lid will pop off and fly into the air at warp speed.

IMPORTANT:
This experiment requires you to wear protective safety glasses.

WARNING!
It's impossible to do this activity just once. It is addicting and habit-forming. Proceed at your own risk! You've been warned.

7. You should have two pieces of Alka-Seltzer tablet left. Repeat the experiment using one of the pieces of Alka-Seltzer, but this time you decide on the amount of water to put in the film canister. Do you think that will make any difference?

8. Use the last piece of Alka-Seltzer to make up your own experiment. What do you want to find out? How are you going to do it? What are you going to measure?

9. Go ahead and experiment!

TAKE IT FURTHER

If you have another Alka-Seltzer tablet, divide it into four equal pieces. This time you're going to determine if changing the temperature of the water has any effect on the speed of the reaction. Repeat the same procedure as before, but change the temperature of the water in each of the four trials and write down your observations. You may need to experiment with several different film canisters until you find one that really pops. It's important that the film canister has a tight seal or it won't pop very well.

For a real eye-popping demonstration, fill ten or more film canisters one-half full with water and drop small pieces of Alka-Seltzer into each one. Quickly put the lids on the canisters and stand back. Popcorn, anyone?

Alka-Seltzer Rocket

Here's a clever variation of the classic Pop Your Top activity, but this time you launch the bottom of the film canister like a rocket. 3-2-1, blast off!

1. Start by sealing the end of the cardboard tube with several pieces of duct tape or use a plastic tube with one end sealed.

2. Divide an Alka-Seltzer tablet into four equal pieces.

3. Fill the film canister one-half full with water. *Note: Steps four through six have to take place very quickly or the rocket will blast off before you're ready. Read the next few steps first to make sure you understand what is going to happen.*

4. Place one of the pieces of Alka-Seltzer tablet in the film canister and quickly snap the lid on the container.

5. Turn the film canister upside down and slide it (lid first) into the tube.

6. Point the open end of the tube AWAY from yourself and others and wait for the pop. Instead of the lid flying off, the bottom of the film canister shoots out of the tube and flies across the room. Listen carefully and you'll hear people yelling, "Do it again!"

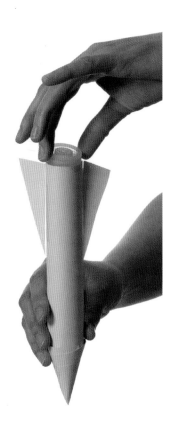

Once you've mastered the technique, it's time to measure how far the film canister rocket flies across the room. After each trial, write down the amount of water you used in the film canister, the size of the piece of Alka-Seltzer (this should not change), and the distance the film canister traveled. What amount of water mixed with a quarter piece of Alka-Seltzer tablet produces the best rocket fuel? Hmm . . . sounds like a good science fair project!

After you've determined the best amount of water to use, try changing the temperature of the water. How does the temperature affect the speed of the reaction? Does warmer or colder water change the distance that the film canister travels?

If you're really creative, you can use construction paper to turn the bottom part of the film canister into a rocket. Wrap some paper around the canister, add some fins, top the whole thing off with a nose cone, and you've got an Alka-Seltzer powered rocket.

WHAT'S GOING ON HERE?

The secret is actually hiding in the bubbles that you observed. The fizzing you see when you drop an Alka-Seltzer tablet in water is the same sort of fizzing that you see when you mix baking soda and vinegar. If you look at the ingredients of Alka-Seltzer, you will find that it contains citric acid and sodium bicarbonate (baking soda). When you drop the tablet in water, the acid and the baking soda react to produce bubbles of carbon dioxide gas.

Carbon dioxide gas builds up so much pressure inside the closed film canister that the lid pops off. The lid is the path of least resistance for the gas pressure building up inside, so it pops off instead of the stronger sides or bottom of the film canister bursting open.

If you tried the experiment again with different temperatures of water, then you also discovered that temperature plays an important part in the reaction. Warm water speeds up the reaction, while colder water takes longer to build up enough pressure to pop off the lid.

We can thank Sir Isaac Newton for what happens next. When the buildup of carbon dioxide gas is too great and the lid pops off, **Newton's Third Law** explains why the film canister flies across the room: for every action there is an equal and opposite reaction. The lid goes one way and the film canister shoots out of the tube in the opposite direction.

BOUNCING SMOKE **BUBBLES**

There's something magical about a bubble. It's just a little puff of air trapped in a thin film of soap and water, but its precise spherical shape and beautiful swirling colors make it a true wonder of science. Bubbles are cool, but bubbles filled with fog are even cooler. Just imagine the cool factor going up tenfold if you could bounce and play with these bubbles. "Boo Bubbles" are what you get when you fill a bubble with a ghostly carbon dioxide cloud. But Boo Bubbles are truly magical because you can roll them on your hands, bounce them off your sleeve, and pop them to release the burst of fog. It's the combination of science and performance art that will have everyone (even you) ooohing and ahhhing.

WHAT YOU NEED

Boo Bubbles™ Kit (available at www.SteveSpanglerScience.com)
Or
Safety glasses
Knit gloves
Gallon-sized plastic jar
3-foot piece of rubber tubing
Liquid soap (Dawn works best.)
Small plastic container

Dry ice

Thick gloves

Bath towel

Making the Dry Ice Bubble Generator

The Dry Ice Bubble Generator pictured below is available as a kit (called the Boo Bubbles Kit) from www.SteveSpanglerScience.com. It's a no-hassle option for the person who wants to get started immediately. It's also possible to make your own Dry Ice Bubble Generator using items that are commonly found at a department store or the plumbing aisle of your favorite hardware store.

You'll need a gallon-sized plastic jar with a 3-foot long piece of rubber tubing attached to the side. The goal is to attach the tubing to the top part of the jar so that the fog created by mixing dry ice and water blows out of the tube when you cover the top of the jar with the lid. The free end of the rubber tubing is attached to a small funnel or something similar to help blow bubbles when it's dipped into a soapy water solution.

The best approach is to start with the plastic jar and spend some time walking through the plumbing aisle of your local hardware store to consider all of the ways to attach a piece of plastic tubing to the jar. It could be as simple as drilling a hole in the jar and attaching the hose with a piece of tape or a dab of caulking or glue. The design is up to you, but be sure to take this book with you so you're prepared when someone asks, "How can I help you?"

WARNING!

Never trap dry ice in a jar without a vent. In other words, there MUST be a hole in the jar to allow the pressure to escape. Otherwise, the pressure will build up and the jar will explode! This could cause serious harm to you or to someone else.

NOTE: You'll need some thick gloves to handle the dry ice. The knit gloves used later in the activity do not provide enough protection for your hands. Find a good pair of leather gloves to protect your hands against the cold temperature of the dry ice and you're set.

LET'S TRY IT!

1. Start by putting on your attractive safety glasses and thick leather gloves. You might need to use a hammer to break up the dry ice into pieces that will easily fit into the jar.

2. Fill the jar one-half full with warm water. Dry ice produces the best fog when you use warm water. Attach the rubber hose to the side of the jar (if it's not already attached).

3. Drop a few good-sized pieces of dry ice into the jar. Immediately, the fog will roll out of the jar. Practice covering the top of the jar with the lid to control the flow of fog out of the tube. You don't have to screw the lid onto the jar. Just hold it on top of the jar to force more or less fog through the rubber tubing.

4. Make a soapy solution by mixing a squirt (that's a *very* technical term!) of liquid soap with about 4 ounces of water in the small plastic container.

5. Dip the free end of the rubber tubing (either the "naked" tubing or the end of the tube "dressed up" with a small plastic cup, funnel, or fitting from the hardware store) into the bubble solution to wet the end of the tube. Remove the tube from the bubble solution with one hand while covering the jar with the lid in the other hand. This will take a little practice, but it's easy once you get the hang of it. The goal is to blow a bubble filled with fog.

6. When the bubble reaches the perfect size, gently shake it off of the tubing and it will quickly fall to the ground (it's heavier than a normal bubble because it's filled with carbon dioxide gas and water vapor). When the bubble hits the ground, it bursts, and the cloud of fog erupts from the bubble. Very cool!

Bouncing Boo Bubbles

This Boo Bubble variation happened accidentally and now it's a must-do whenever you play with Boo Bubbles. A bath towel was stretched out on the table in an effort to make the soapy cleanup just a little easier. To everyone's amazement, some of the fog-filled bubbles bounced on the towel and didn't pop! It just goes to show you that what some people call "play time" is actually high-level research (okay, maybe it's not real *research*, but it is play with a purpose). It's important to mention that not all types of fabric behave the same way. Ponder that for a few minutes before reading about the science of the bouncing bubble in the following pages.

Touchable Boo Bubbles

If fog-filled bubbles will bounce off of a towel, what would happen if you wrapped your hands in fabric and tried to touch or play with the bubbles? You can easily find out by purchasing a pair of inexpensive children's winter gloves. Blow a bubble about the size of a baseball with the Dry Ice Bubble Generator. Bounce the bubble off of your gloves. Try bouncing the bubble off of your shirt or pants. Again, some fabrics work better than others. Try bouncing bubbles on a hand towel or start up a game of volleyball bubbles with another friend who has too much time on her hands.

Giant Boo Bubbles

Regular-sized Boo Bubbles are awesome, but Giant Boo Bubbles are even more awesome! All you need are a few parts and pieces from around the house and you'll be making these giant, fog-filled bubbles in no time. Cover a clean table surface with a thin layer of soap bubble solution and spread it around. Fill the large water bottle with warm water and drop in a few big pieces of dry ice. Again, *NEVER put any type of lid on the bottle or do anything that would seal the bottle closed. The rapidly expanding gas could result in an explosion.*

Hold the large plastic hose (similar to the kind you'd find on a shop vacuum) over the top of the large water bottle. The carbon dioxide cloud will start flowing out of the hose. Make sure you don't plug the hose so the gas can't escape. That never ends well . . . trust us.

Dip the open end of the hose into the bubble solution and put it down on the soap-covered table. A giant Boo Bubble will start forming on the surface of the table! Keep the nozzle down and your bubble will just get bigger and bigger and bigger. When the bubble finally pops, all of that carbon dioxide gas will escape, leaving a ghostly fog behind.

While blowing bubbles indoors, you might have noticed the occasional bubble that fell to the carpet but didn't pop. Regular bubbles burst when they come in contact with just about anything. Why? A bubble's worst enemies are oil and dirt. Soap bubbles will bounce off of a surface if it is free of oil or dirt particles that would normally puncture the soap film. They break when they hit the ground, but they don't break if they land on a softer fabric like gloves or a towel.

Dry ice is frozen carbon dioxide (CO_2) Under normal atmospheric conditions, CO_2 is a gas. Only about 0.035% of our atmosphere is made up of carbon dioxide. Most of the air we breathe is nitrogen (79%) and oxygen (20%). Instead of melting, dry ice turns directly into CO_2 gas. It does not melt like real ice because it skips the liquid stage and goes straight from solid to gas. When you drop a piece of dry ice in a bucket of water, the gas that you see is a combination of carbon dioxide and water vapor. So, the gas is actually a cloud of tiny water droplets.

Dry ice must be handled with care because it is -109.3°F (-78.5°C). It must be handled using gloves or tongs—otherwise, it will cause severe burns if it comes in contact with your skin. Never put dry ice into your mouth!

Grocery stores use dry ice to keep food cold during shipping. Some grocery stores and ice cream shops will sell dry ice to the public (especially around Halloween) for approximately $1 per pound. Dry ice comes in flat square slabs a few inches thick or as cylinders that are about 3 inches long and about half an inch in diameter. Either size will work fine for your dry ice experiments. Remember the science . . . dry ice turns directly from a solid into a gas—a process called **sublimation**. In other words, the dry ice in the grocery bag will literally vanish in about a day! The experts tell us that dry ice will sublimate (turn from a solid into a gas) at a rate of 5 to 10 pounds every 24 hours in a typical vented ice chest. It's best to purchase the dry ice as close to the time you need it as possible. This is the one time when last-minute shopping is necessary.

COLORFUL
CONVECTION CURRENTS

Cool, crisp, clean mountain air has long been an important reason why people move to Colorado. Unfortunately, the air in the city of Denver isn't quite so clean. By the 1970s, the pollution hanging over the city had a name—the "brown cloud." Denver's location at the foot of the Rocky Mountains makes it prone to temperature inversions in which warm air higher up traps cooler air near the ground, preventing pollutants from rising and escaping into the atmosphere. However, the phenomenon of temperature inversion is not unique to Denver, as evidenced by numerous reports of similar brown cloud sightings over major cities throughout the world. This demonstration provides a great illustration of what's really happening in the atmosphere as hot and cold air masses meet.

WHAT YOU NEED

Four empty identical bottles (Browse the juice aisle at the grocery store to find bottles similar to those pictured in the book. As a general rule of thumb, the mouth of the bottle should be at least an inch in diameter.)

Access to warm and cold water

Food coloring (yellow and blue)

3 x 5-inch index card or an old playing card

Masking tape

Pen

Paper towels

LET'S TRY IT!

1. Fill two of the bottles with warm water from the tap and the other two bottles with cold water. Use masking tape and a pen to label the bottles with the words "HOT" and "COLD."

2. Use food coloring to color the warm water yellow and the cold water blue. Each bottle must be filled to the brim with water.

3. The next step can be a little tricky, but with practice you'll have no problem. Your first observation will be what happens if the bottle with warm water rests on top of the bottle filled with cold water. To accomplish this, place the index card or old playing card over the mouth of one of the warm water bottles (remember, it's the bottle with the yellow water). Hold the card in place as you turn the bottle upside down and rest it on top of one of the cold water bottles. The bottles should be positioned so that they are mouth to mouth with the card separating the two liquids. Just make sure to have towels close by in case everything doesn't go exactly as planned.

4. Carefully slip the card out from in between the two bottles, making sure that you are holding onto the top bottle when you remove the card. Take a look at what happens to the colored liquids in the two bottles.

5. For the second variation of the experiment, you need to have the warm water on the bottom and the cold water on top. Repeat steps three and four, but this time place the bottle of cold water on top of the warm water. Carefully remove the card and watch what happens.

WHAT'S GOING ON HERE?

Hot air balloons rise because warm air is lighter and less dense than cold air is. Similarly, warm water is lighter in weight or less dense than cold water is. When the bottle of warm water is placed on top of the cold water, the more dense cold water stays in the bottom bottle and the less dense warm water is confined to the top bottle. However, when the cold water bottle rests on top of the warm water, the less dense warm water rises to the top bottle and the cold water sinks. The movement of the water is clearly seen as the yellow and blue food coloring mix, creating a green liquid.

The movement of the warm and cold water inside the bottles is referred to as a **convection current**. In our daily life, warm currents can occur in oceans, like the warm Gulf Stream moving up north along the American Eastern Seaboard. Convection currents in the atmosphere are responsible for the formation of thunderstorms as the warm and cold air masses collide.

Although the bottles whose colored liquids mix are more interesting to watch, the other set of warm and cold water bottles helps to illustrate another important phenomenon that occurs in the atmosphere during the winter months. During daylight hours, the Sun heats the surface of the Earth and the layer of air closest to the Earth. This warm air rises and mixes with other atmospheric gases. When the Sun goes down, the less dense warm air high up in the atmosphere often blankets the colder air that rests closer to the surface of the Earth. Because the colder air is more dense than the warm air, the colder air is trapped close to the Earth and the atmospheric gases do not mix. This is commonly referred to as **temperature inversion**.

What are the results of temperature inversion? Air pollution is more noticeable during a temperature inversion since pollutants such as car exhaust are trapped in the layer of cool air close to the Earth. As a result, state agencies in many parts of the country oxygenate automobile fuels during winter months with additives like MTBE in an attempt to reduce the harmful effects of trapped pollution. This trapped pollution is what causes the brown cloud effect. Wind or precipitation can help alleviate the brown cloud effect by stirring up and breaking up the layer of warm air that traps the cold air and pollution down near the surface of the Earth.

WALKING ON EGGSHELLS

The phrase "walking on eggshells" is an idiom that is often used to describe a situation in which people must tread lightly around a sensitive topic for fear of offending someone or creating a volatile situation. *Literally* walking on eggshells would require exceptional caution, incredible skill, and a sense of self-control that would be nothing short of amazing. Wait just a second . . . what if eggs were really much stronger than most of us imagine? What if nature's design of the incredible edible egg was so perfect that the thin, white outer coating of an egg was strong enough to withstand the weight of your body? Wake the kids! Phone the neighbors! It's time for the Walking on Eggshells challenge.

WHAT YOU NEED

A few dozen eggs that are in egg cartons (Select large-sized eggs.)

Large plastic trash bag

Bucket of soap and water (and some disinfectant)

Barefoot friends

NOTE: There's a very high probability that you'll break a few eggs while attempting to learn this amazing trick. Since raw eggs carry the danger of Salmonella, it's important that you clean up, wash your hands, and disinfect the area. Even if you don't break an egg, it's still a good idea to wash your hands (and feet!) after handling eggs.

LET'S TRY IT!

Here are a few demonstrations that are a good warm-up for the main event.

Squeeze an Egg Without Breaking It

1. Start by covering your demonstration area with the large plastic trash bag. This will make the cleanup much easier (if you happen to make a mess, that is).

2. Remove any jewelry you might be wearing on your fingers. You'll need a completely naked hand to attempt this feat.

3. Place the raw egg in the palm of your hand. Close your hand so that your fingers are completely wrapped around the egg.

4. Squeeze the egg by applying even pressure all around the shell. Get past your fear of breaking the egg and really put the squeeze on it!

5. To everyone's amazement (mostly your own) the egg will not break. If you had been wearing a ring, the uneven pressure of the band against the shell would have most likely resulted in an amusing display of flying egg yolk. Fun to watch but a bummer to clean up.

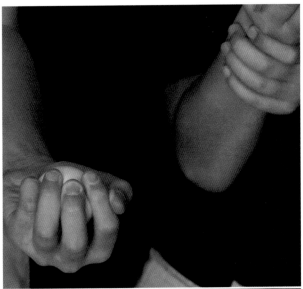

6. Never rest on your initial success! Try changing the way you hold the egg in your hand to see if the orientation of the egg makes any difference when you put the squeeze on it. If you're really brave, hold the egg in the palm of your left hand and apply pressure to the egg with your right hand. Consider it an unusual way to "egg-press" yourself. Now, where are those towels you meant to set out before you started all of this craziness?

Getting Past the Fear Factor

If you're a little nervous about the outcome, try sealing the raw egg in a ziplock plastic bag before putting the squeeze on it. Or, hold the egg over the sink if you're in the super-brave category.

The Main Event—Walking on Eggshells

Eggs are amazingly strong, despite their reputation for being so fragile. But, are they strong enough to support the weight of your body?

1. If you just want to attempt the feat of standing on eggs, you'll only need two cartons of eggs (two dozen eggs). If, however, you're feeling up to the Walking on Eggshells challenge, pick up six or eight cartons of large-sized eggs.

2. Spread the plastic trash bag (or bags) out on the floor and arrange the egg cartons into two rows.

3. Inspect all of the eggs to make sure there are no breaks or fractures in any of the eggshells. Make any replacements that might be necessary.

4. It's important to make sure all of the eggs are oriented the same way in the cartons too. One end of the egg is more "pointy" while the other end is more round. Just make sure that all of the eggs are oriented in the same direction. By doing this, your foot will have a more level surface on which to stand.

5. Remove your shoes and socks . . . and pick the lint out from between your toes. (This has no bearing on the success of the challenge, but let's face it, toe fuzz is kind of gross.)

6. Find a friend to assist you as you step up onto the first carton of eggs. The key is to make your foot as flat as possible in order to distribute your weight evenly across the tops of the eggs. If the ball of your foot is large, you might try positioning it between two rows of eggs instead of resting it on the top of an egg.

7. When your foot is properly positioned, slowly shift all of your weight onto the egg-leg as you position your other foot on top of the second carton of eggs.

8. There will be creaking sounds coming from the egg carton, but don't get nervous. Ask your friend to step away and allow your fans to click pictures. Just think . . . for all the *right* reasons, you'll be an Internet sensation in just minutes.

9. If you have more than two cartons of eggs, what are you waiting for? Keep walking! The cheers and wild screams from your fans grow louder with each step you take until finally you land on firm ground and marvel at your success.

Okay, there's a second scenario that we should mention: you forget to make your foot as flat as possible, your friend doesn't provide any support, and your foot crushes through eight of the twelve eggs. As the goo erupts from between your toes, you think to yourself, "Maybe the other carton will be better." Quickly you discover that both feet are covered in eggy goo and the experiment is a complete failure. Don't worry, your fans are still taking pictures and you're still going to be an Internet sensation, but for a completely different reason. Ah, show business!

TAKE IT FURTHER

Instead of standing on the eggs, place a board or tile on top of a dozen eggs and test their strength by stacking books one at a time on top of the eggs.

Upside-down plastic soda bottle caps can be used in place of an egg carton to keep the eggs in an upright position while you're attempting your strength test. Try arranging the eggs into an "X" pattern to fully support the board. How much weight will five eggs support before cracking under the pressure?

WHAT'S GOING ON HERE?

Plain and simple, the shape of the egg is the secret! The egg's unique shape gives it tremendous strength, despite its seeming fragility. Eggs are similar in shape to a three-dimensional arch, one of the strongest architectural forms. The egg is the strongest at the top and the bottom (or at the highest point of the arch). That's why the egg doesn't break when you add pressure to both ends. The curved form of the shell also distributes pressure evenly all over the shell rather than concentrating it at any one point. By completely surrounding the egg with your hand, the pressure you apply by squeezing is distributed evenly all over the egg. However, eggs do not stand up well to uneven forces, which is why they crack easily on the side of a bowl (or why it cracked when you just pushed on one side). This also explains how a hen can sit on an egg and not break it, but a tiny little chick can break through the eggshell. The weight of the hen is evenly distributed over the egg, while the pecking of the chick is an uneven force directed at just one spot on the egg.

If you guessed that the egg carton probably played a role in keeping the eggs from breaking, you're right. Joseph Coyle is credited as the inventor of the first container made specifically to keep eggs from breaking as they were transported from the local farm to the store. As the story goes, Coyle invented the egg carton in 1911 as a way to solve a dispute between a farmer and a hotel operator who blamed the farmer for delivering broken eggs. Coyle designed a container made out of thick paper with individual divots that supported each egg from the bottom while keeping the eggs separated from one another. As legend has it, the fully loaded egg carton can even be dropped, and if it lands just right, the eggs will survive the fall.

SKATEBOARD **ROCKET CAR**

Anyone who has tried it will tell you that the Mentos and Diet Coke geyser is totally awesome! Here's a way to turn the explosive geyser on its side and transform it into a horizontal rocket motor that powers a skateboard over the ground.

WARNING! Adult supervision is required because there's going to be soda flying everywhere. If you're an adult doing this, the good news is that diet soda isn't sticky, and it's easy to get off your neighbor's car.

LET'S TRY IT!

1. Find a safe and suitable launch site for your Skateboard Rocket Car. Empty paved parking lots or paved driveways work as great launch areas.

2. You'll need to attach the full soda bottle to the skateboard so that the geyser has a clear shot *backward*. Make sure there is plenty of room *behind* the geyser because when it launches, it will really take off!

3. Check out the design of your skateboard. If the skateboard tips curve up on the ends, the geyser might be deflected and the skateboard won't move much. You may need to add some layers of cardboard between the soda bottle and the skateboard so the rocket "exhaust" clears the tip of the skateboard.

4. When you've got the bottle placed just the way you want it, have your helper hold it tightly as you wrap the duct tape once around the lower end of the bottle and the skateboard and then once around the top end of the bottle and the skateboard near its tip. Now that the bottle and the board are joined together, you've made a launching system!

5. Put the system on the ground and check the alignment of the bottle with the skateboard. It should be straight and centered, and the open end of the bottle needs to be above the curved tips of the skateboard. When you're happy with the placement of the system, wrap two more layers of duct tape around each end of the system right on top of the first layer. Any color of duct tape will do, but even scientists need to let their inner expressive selves loose sometimes. Be daring!

6. Hold the system off of the ground, keeping it vertical, with the top of the bottle pointing up. Remove the bottle cap. Now you're getting serious!

7. Ask your helper to open one end of the Mentos package and loosen the mints a little so they can slide out easily. Have your helper hold the mints in one hand and place the open end of the Mentos package above the opening of the bottle (or trade places and *you* do it). Then, lift the package straight up and slide all of the Mentos mints into the soda all at once with the other hand. Do this by pinching the closed end of the wrapper between your fingers and sliding the mints toward the open end. This will push the Mentos quickly into the bottle in one continuous motion. ***Note: It is essential, but tricky, to drop the mints into the bottle all at the same time. If they don't fall into the bottle at the same time, the reaction will start before you're ready and your Skateboard Rocket Car may take off too soon! Skip ahead to the "Take It Further" section to solve the problem using Steve Spangler's Geyser Tube® launcher.***

8. There's only a split second to set the system on the ground before the geyser erupts and shoves the Skateboard Rocket Car downrange. Once the Mentos drop into the bottle, place the Skateboard Rocket Car on the ground as fast as you can and get out of the way! If you hesitate, you'll get a tasty but very messy shower and then, aw shucks, you'll have to do it again.

The Skateboard Rocket Car is an activity you (and your now diet soda–soaked adult helper) will want to do over and over. Perform the activity again using a different variable—a different brand of diet soda, a different type of soda (regular instead of diet), a different sized bottle of soda, a different type of mint, or a different skateboard. The options are endless, so keep experimenting, all in the name of science . . . or just because it's really fun!

To make the launch sequence simpler, cleaner, and more consistent, you could use Steve Spangler's Geyser Tube® launcher (available at www.SteveSpanglerScience.com). With this device, dropping the Mentos into the bottle is the same every time, and its design also intensifies the blast. The designers of the Geyser Tube shared this secret: this special device does more than just drop the mints into the soda. The special design utilizes the Venturi effect to maximize the eruption of the geyser. The **Venturi effect** is similar to placing your thumb over the end of a garden hose that's shooting out water. The velocity of the water outside the hose will increase when you decrease the size of the opening. The same thing happens when you use the Geyser Tube on the Skateboard Rocket Car.

You can compare the performance of your Skateboard Rocket Car with and without using the Geyser Tube. Use chalk or tape to mark the starting line on your driveway and then launch one original Skateboard Rocket Car and one *turbo* Skateboard Rocket Car (using the Geyser Tube). Make sure you keep all other variables the same. The only difference in your Skateboard Rocket Cars should be the use of the Geyser Tube. Measure the distance of the launches and compare the data. Chart your data, write up your discoveries, and you have an "explosive" and awesome science fair project!

WHAT'S GOING ON HERE?

Soda pop basically consists of lots of sugar (sucrose, fructose, or a diet sweetener), some flavoring, water, and preservatives. What gives soda its bubbly appeal is invisible carbon dioxide gas (CO_2), which is forced into the liquid using tons of pressure. Until you open a soda, which makes lots of room, the gas mostly stays suspended in the liquid and can't collect to form bubbles, which is what gases naturally do. Even when the bottle is open, however, most of the gas stays in the liquid and provides the fizziness of the soda. However, if you shake the soda and then open it, the gas is quickly released from the protective hold of the water molecules and escapes with a *whoosh*, taking some of the liquid along with it.

There are other ways to cause the gas to escape. Just drop something into a glass of soda and notice how bubbles immediately form on the surface of the object. For example, adding salt to soda causes it to foam up because thousands of little bubbles form on the surface of each grain of salt. Also, compare the foaming of diet soda to regular soda when each is poured over ice.

The Mentos Secret

The reason why Mentos mints work so well is twofold—tiny, *very* tiny, pits on the surface of the mint and the weight of the mints themselves. Each mint has thousands of micro-pits all over its surface. These tiny pits are called **nucleation sites**, and they're perfect places for carbon dioxide bubbles to form. As soon as the Mentos hit the soda, bubbles form all over the surfaces of the mints and then quickly rise to the surface of the liquid. Couple this with the fact that the mints are heavy and sink to the bottom of the bottle and you've got a double whammy. The gas released by the Mentos literally pushes all of the liquid up and out of the bottle in an incredible soda blast. The geyser continues to erupt as long as the pits remain on the surface of the mints. Eventually, enough of the surface is dissolved so that it becomes too smooth for the gas to rapidly collect. At that point, the reaction slows and stops.

Use a magnifying glass to look closely at the surface of a single Mentos mint and compare it to one that has been dropped into diet soda. You might be able to see the tiny pits on the surface of the new mint. The tiny pits are the key to all of the soda geysers you're going to make now and at every party you go to in the future. You'll be the most popular person at the party!

Physics in Motion

The Skateboard Rocket Car is also a great example of physics in motion. Sir Isaac Newton, an English scientist, writer, philosopher, mathematician, and more, discovered some laws of physics way back in the late 1600s and early 1700s. In a word, he was *really* smart! In order for anything to move, it must obey **Newton's Laws of Motion**, and there are three of them.

Simply stated, a Skateboard Rocket Car standing still will remain standing still; a moving Skateboard Rocket Car will continue moving in a straight line. These conditions don't change unless an outside force is strong enough to make the system move faster or slower, stop, or change directions (that is **Newton's First Law of Motion**). Not too complicated, right?

Newton's Second Law says that to cause the system to move faster or slower, stop, or change directions, the force used has to overcome the inertia (or motion) the Skateboard Rocket Car already has. Okay, this is getting a little more complicated.

And for the big finale, **Newton's Third Law of Motion** states that for every action there is an equal and opposite reaction. The force of the soda geyser whooshing backward out the bottle is exactly matched by a force pushing the Skateboard Rocket Car forward. The stronger the backward geyser, the faster the car moves forward—that is why the Geyser Tube creates a more powerful launch.

Who knew physics could be so much fun?

SCIENCE IN **MOTION**

BALANCING **NAILS**

The object is to balance a bunch of nails on the head of a single nail. All of the nails have to be balanced at the same time and cannot touch anything but the top of the nail that is stuck in the base. Are you up to the challenge?

WHAT YOU NEED

Block of wood (4 inches square and about ½ inch thick)

12 identical nails with heads (The nails should be 10-penny size or larger.)

Hammer

LET'S TRY IT!

1. Hammer one of the nails into the center of the block of wood. It's a good idea to measure and predrill the hole to avoid splitting the wooden block. It's important that this nail be standing as straight as possible.

2. Place the wood block flat on a desk or table. The challenge is to balance all of the nails on the standing nail in the wooden block. To win the challenge, none of the 11 nails may touch the wood block, the desk or table, or anything else that might help hold them up. No additional equipment other than the wood block and the nails may be used.

3. Need help? The trick to balancing the nails has to do with their center of gravity or balancing point. Lay one nail on a flat surface and place the other nails across this nail, head to head as shown in the photograph on the following page. Finally, place another nail on top of this assembly, head to tail with the second nail.

4. Carefully pick up the assembly and balance it on the upright nail. Voilà!

TAKE IT FURTHER

Slowly remove one nail at a time. How many nails can you remove before the system collapses? Which nails are necessary for the system to remain in balance?

If you're really ambitious, you can try your luck at our large-scale version using landscape nails (often called "spikes") and a friend as the base. Have a friend hold one of the large nails while you attempt to balance the other 11 nails on the head of the single nail. Here's a bit of advice—start with the small-scale version and work your way up.

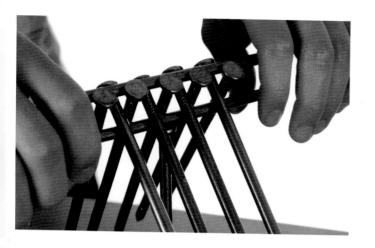

Gravity pulls any object toward the center of the Earth as if all of its weight were concentrated at one point. That point is called the **center of gravity**. Objects fall over when their center of gravity is not supported. For symmetrical objects like a ball or a meter stick, the center of gravity is exactly in the middle of the object. For objects that are not symmetrical, like a baseball bat, the center of gravity is closer to the heavier end.

The stability of the nails depends on their center of gravity being right at or directly below the point where they rest on the bottom nail. Add too many nails to the left or right and they become unstable and fall off.

This "scientific" puzzle is trickier than it looks. The best way to solve it is to think of an idea and then try it out. Even if it doesn't work, you might think of another idea at the same time. The key is to not get frustrated and give up. Keep trying. You might even have to sleep on an idea and come back to it the next day. You may want to share your ideas with others to see if they have a different approach to solving the problem. This problem-solving process is exactly like the scientific method—ask a question, run some tests, ask another question, run some more tests, and eventually come to a conclusion. If your experiment or "solution" doesn't work, that's okay. Some of the greatest scientific discoveries have been made by mistake!

CORK IN A WINE BOTTLE **PUZZLE**

If a cork falls into the bottom of a wine bottle (accidentally or on purpose), how can you get it out without ruining the cork, the bottle, or both? It's a puzzle that seems challenging, but actually it can be solved with a simple science secret.

WHAT YOU NEED

Empty wine bottle

Cork

Rubber mallet

Small stick

Handkerchief or cloth napkin

LET'S TRY IT!

1. If you have an empty wine bottle, a cork, and a handkerchief lying around the house, you have the basic materials you need to attempt this challenge.

2. Put the cork back in the wine bottle. This can be a bit tricky, so push it into the top of the bottle and then use a rubber mallet to push it as far into the bottle as you can. You could also turn the bottle over and tap the cork lightly on the ground until the cork is even with the top of the bottle.

3. Use a small stick to push the cork all the way into the bottom of the bottle. Now you have what some people call an "Impossible Bottle"—a bottle that contains something that appears to be too large to fit through the neck of the bottle and that won't easily come out. Some Impossible Bottles feature ships, decks of cards, tennis balls, scissors, knotted ropes, and many other unusual items inside them.

4. Call your friend over and present him with this challenge: "Can you pull the cork out of the bottle without destroying it or the bottle?" Remind him that he can't set the cork on fire or put something down into the bottle that will break the cork up into small pieces or break the bottle. Let him ponder the challenge for a while. Maybe he'll consider putting water and Alka-Seltzer down into the bottle, building up the pressure, and exploding the cork out of the bottle. Or maybe he'll pour some liquid nitrogen down into the bottle and douse it with warm water, causing a huge cloud (and the cork) to erupt out of the bottle. (These techniques are not recommended and could cause damage to the cork, the bottle, or your friend!)

5. When your friend gives up and begs you to explain how to get the cork out of the bottle, here is one solution you might share

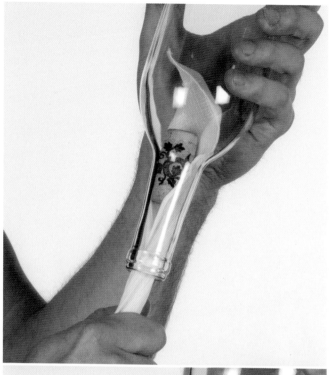

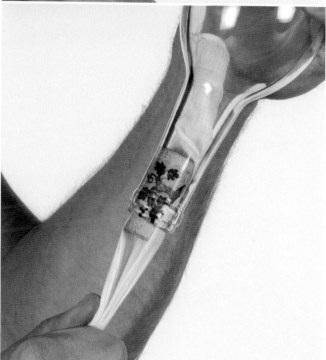

with him. Push the handkerchief down into the bottle, leaving about half of the handkerchief sticking out of the bottle so that you can grab onto it.

6. Tip the bottle upside down and gently shake it until the cork is lodged between the handkerchief and the inside of the bottle near the neck of the bottle.

7. Grab the dangling end of the handkerchief and slowly pull the handkerchief out of the bottle. The handkerchief is slippery and so doesn't provide the normal friction between the cork and the bottle. If you continue to pull on the handkerchief, *BAM!*—the cork pops right out of the bottle.

TAKE IT FURTHER

See if you can remove a cork from an empty wine bottle using a produce bag from the grocery store instead of a napkin or handkerchief. What other items could you use to try to get the cork out of the bottle?

WHAT'S GOING ON HERE?

There's really no fancy scientific explanation needed to describe how the cork comes out of the bottle. It boils down to just the simple science of **friction**. Wine manufacturers use corks in bottles because the friction between the cork and the glass forms a nice tight seal. If you remove the friction (by using the handkerchief or a plastic bag), you can very easily pull the cork back out of the bottle. How's that for a science mind-bender?

DRIPPING **CANDLE SEESAW**

This scientific amusement is easily overlooked when you're thumbing through the pages of an old science book. How exciting can a few candles attached to a stick really be? Don't overlook one of true treasures of this book. The updated design makes this seesaw candle experiment a must-see part of your growing repertoire.

WHAT YOU NEED

Pointed-tip scissors

2 small birthday candles

3 small paper clips

Ruler

Small plastic cup

Thumbtack

2 straws

Matches or lighter

Newspaper or surface you don't mind getting wax on

WARNING! Since flames are involved, young scientists will need to round up some adult supervision before attempting this experiment.

LET'S TRY IT!

1. Using the pointed end of the scissors, poke a hole into the center of the bottom of the small plastic cup. The hole should be just big enough to allow the end of one of the straws to slide through it.

2. When you have the hole in the cup, slide the end of a straw through the hole. Turn the cup upside down and place it so that it is flat and sturdy on the table.

3. Using the ruler, find the middle of the remaining straw. Once you have found the middle of the straw, poke a hole all the way through it and out the other side using the thumbtack. Try to keep the thumbtack level as you poke through the straw.

4. Take one of the paper clips and straighten out the smaller loop so that it is almost straight. Take the end of the straightened side and bend it upward. The shape you end up with should look like an *L* connected with a *J*.

5. Slide the *L* end of the bent paper clip through the straw where you punched the holes. Slide the paper clip so that the straw is at the bottom of the *L*.

6. Now take the *J* side of the bent paper clip and put it into the top of the straw opposite the cup. Your apparatus should look like a seesaw now.

7. Insert the flat "wickless" end of each candle into the two ends of the seesaw straw. To hold each candle in place, slide a paper clip across the face of the straw, enclosing both it and the candle.

8. Balance the seesaw by sliding the two candles in or out of the straw.

9. Rest the entire setup on a piece of newspaper to catch the hot wax drippings.

10. Make sure the seesaw is balanced and at rest when you pose the first question: "What will happen when I light one of the candles?" Give every person a chance to share his or her predictions. Go ahead and light one of the candles and observe what happens.

11. The next question seems only logical: "What will happen when I light the second candle?" After about 15 seconds, the candles will start to move up and down in a seesaw-like motion. As time goes on, the swing of the straw gets greater and greater until eventually you get a full rotation. Some people will burst out in spontaneous applause while others will stare curiously at the contraption with a puzzled look on their face.

This demonstration is a great conversation starter, so it's important to give people enough time to ponder a solution and share their ideas *before* you share the science secret.

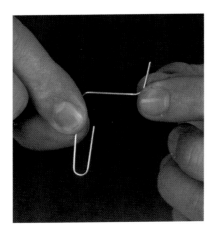

WHAT'S GOING ON HERE?

Let's start by thinking about the seesaw you played on as a young kid. The seesaw is not just a playground toy—it is an example of a simple machine. In physics, simple machines are devices that make it easier to do work. A lever is an example of a simple machine. A lever is a straight rod or board that pivots on a stationary

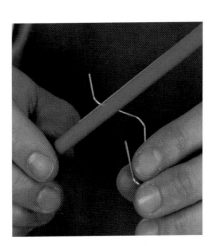

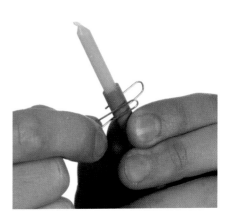

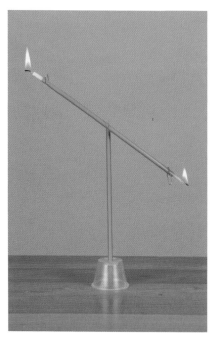

point called a pivot point, or **fulcrum**. Levers are often used to lift heavy loads. A seesaw, a shovel, a fishing pole, a pair of scissors, a baseball bat, and a wheelbarrow are all examples of levers.

What makes a playground seesaw fun is changing the masses (or the heavy loads!) on the ends of the board. However, unlike the Dripping Candle Seesaw, the playground toy moves up and down with some help from the pushing force of little legs. Since the candles don't have legs, there has to be another explanation.

As you might have guessed, part of the secret of the Dripping Candle Seesaw is actually in the name of the activity . . . *dripping*. The rotational action of the Candle Seesaw comes from the changing mass of the candles—**potential energy** that gets turned into rotational kinetic energy.

While it might have just sounded like a polite thing to do, putting down the newspaper to catch the wax drippings is the key to understanding how this works. Over time, you noticed that the candles began to drip wax, and in doing so they lost potential energy. If both candles dripped wax at exactly the same rate, there would be no movement—but this happens only when both candles are lit at exactly the same time. As the heavier candle in the seesaw moves downward, the angled flame causes the candle wax to melt faster and drip more. When the dripping candle loses enough mass, it also loses potential energy and moves upward (just like a seesaw). Now the candle on the other end moves downward, the angled flame melts the wax, which drops onto the paper, and the seesaw is set back into motion.

The careful observer is quick to point out that the height of the swing (otherwise known as the **amplitude**) increases with each cycle. Eventually the **kinetic energy** (or the energy of motion) is great enough to produce a full rotation, and with that comes cheers of excitement (or just a smirky grin followed by the utterance, "That's cool.").

Bob Becker first shared this demonstration with me at a conference for chemistry teachers at Old Dominion University in Norfolk, Virginia. Interestingly enough, Bob pointed out that the demonstration was first published in H. H. Windsor's book entitled *The Boy Mechanic, Book 2: 1000 Things for Boys to Do*, published by Popular Mechanics Press in 1915. While I first want to be quick to apologize for the book's title, I also want to point out that Bob's lesson is an important one to remember—*the old is forever new.*

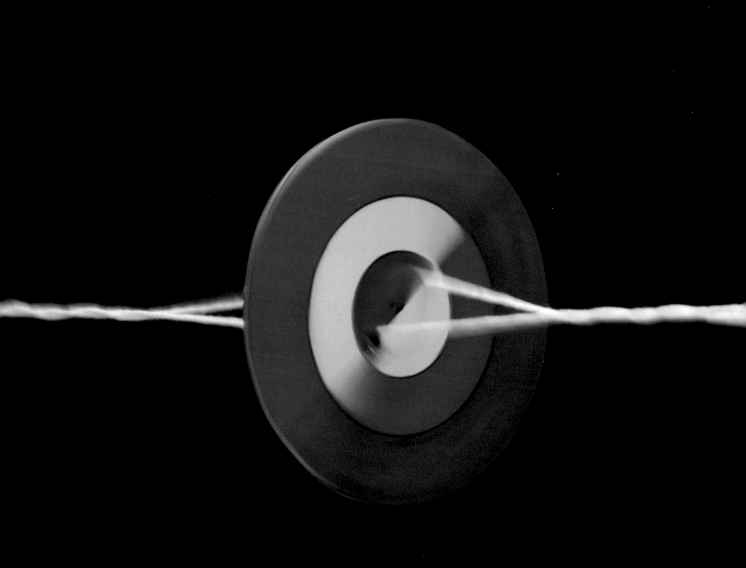

COLOR MIXING **WHEEL**

Here's an amazing way to combine scientific principles of physics with the visual science of color mixing to create a gizmo that will have you twirling and spinning for hours. The activity isn't just visually spectacular, it's scientifically sound and filled with just enough open-ended curiosity that you're bound to make many versions of the Color Mixing Wheel.

WHAT YOU NEED

White corrugated cardboard

Pointed tip scissors

Red, blue, and yellow markers

String or yarn

Large plastic cup

Soda bottle cap

Adult supervision

LET'S TRY IT!

Creating the Color Mixing Wheel

1. The mouth of a plastic cup works great as a template for the circle. Use the cup to trace a circle onto a piece of white corrugated cardboard. Try to get the circle to be between 4 to 6 inches in diameter.

2. Cut out the traced circle.

3. Use the bottom of the plastic cup to trace a smaller sized circle onto the cardboard disc. For the third circle, trace around a plastic soda bottle cap. Of course, you can use anything you want to make the circles, but the goal is to make each circle smaller than the other ones by roughly the same amount (an inch or two). Think of a colorful archery target in which the bands of red and blue and yellow are all about the same width. This will enhance the visual aspect of the experiment.

4. Draw a single line through the middle of the disc that spans the entire diameter of the disc. Each of the three circles in the disc should now be divided in half.

5. Color half of the smallest circle blue and the other half yellow. Color the middle circle half red and half yellow. Finally, color the largest circle half blue and half red.

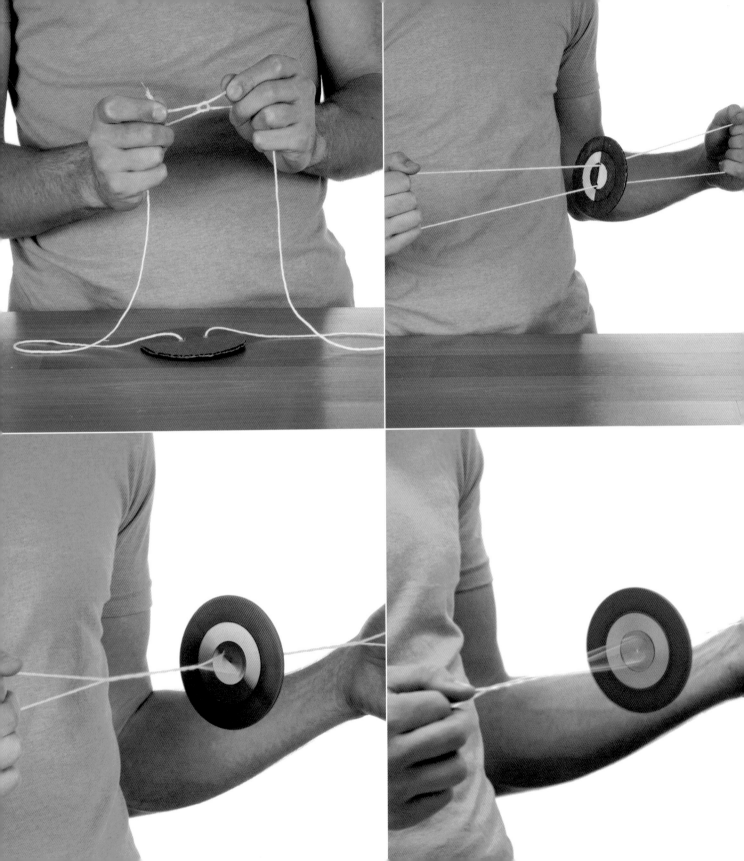

6. Find someone who's good with scissors for the next step. Using the pointed tip of the scissors, place two holes in the cardboard disc. Use a ruler to make sure the holes are an equal distance from the center of the disc and are about 1 inch apart.

7. Use the scissors to cut a piece of string or yarn that is 4 feet long.

8. Thread the string or yarn through each of the holes in the disc and tie the ends of the string together. Make sure the knot you tie is reliable and able to withstand a substantial amount of force. You are going to be tugging pretty hard on it.

Performing the Experiment

1. Start by holding the string on both sides of the disc with your hands. Make sure to slide the disc to as close to the center of the string as possible.

2. Spin the disc in a motion similar to turning a jump rope. This is a quick way to get the string wound up.

3. Once the string on both sides of the disc is twisted, pull the string tight to get the Color Mixing Wheel spinning. It might take a little practice to get it just right.

4. Once you have the hang of how the Color Mixing Wheel works, you'll be able to keep it going as long as you want.

5. You may have noticed that the colors you put on the Color Mixing Wheel were the three primary colors: red, blue, and yellow. Once you start spinning the wheel, what do you notice about each of the three colored circles on the cardboard disc? What do you think makes this happen?

WHAT'S GOING ON HERE?

Let's start with the visual part of the experiment—color mixing. The colors you put on the Color Mixing Wheel are the three primary colors: red, blue, and yellow. When you combine two primary colors, you get the secondary colors: purple, green, and orange. Obviously, the individual colors on the wheel are not mixing. The color mixing that happens is due to the speed at which the wheel is spinning as the string twists. The colors are spinning at such a rate that your brain is unable to process them as the individual colors that are on the wheel. Instead, your brain takes a shortcut and creates the secondary colors.

Now, why does the string continue to twist? The answer lies in physics and, in particular, momentum. Once you have the string twisted, pulling on each end causes it to go tight. When the string is pulled tight, it wants to be completely straight. In going straight, the string unwinds itself and causes the disc to spin in one direction. But the string doesn't stop once it's unwound. It speeds past and gets twisted again in the other direction. The momentum from pulling the string tight keeps the disc spinning until all the momentum is gone. Then you pull the strings tight again and set the disc spinning in another direction.

THE UNBELIEVABLE
PENDULUM CATCH

It's a classic comedy bit from the silent movie era. The villain hoists a piano to a third- or fourth-story window just waiting for the unsuspecting victim to stand below the precariously perched piano. The rope is cut, the piano crashes to the ground . . . but the victim always seems to stumble away in a daze. Okay, that's the movies, but there is something interesting about a free-falling pendulum. So, get a piano, a rope, and an army of friends . . . okay, maybe not. First, try your luck with simple objects you can find around the house.

WHAT YOU NEED

15 identical metal washers

Shoelace, cotton string, or a piece of yarn that is approximately 27 inches long

LET'S TRY IT!

1. To get started, thread your string through 14 of the 15 metal washers.

2. Take the end of the string you just threaded through the metal washers and tie it back onto the string right above the stack of washers. Basically, you are making a small loop of string with washers on it.

3. Thread the other end of the string through the middle of the remaining washer and tie the string so that you have a string with 14 washers at one end and one washer at the other end. The ideal ratio of washers on the ends of the string is 14 to 1, and that's just what you have.

4. With the string-washer apparatus you have constructed, grab the single-washer end with one hand and hang the heavy end of the string over your opposite hand's pointing index finger.

5. Pull the single-washer end of the string so that the 14 washers are touching your pointing index finger. Make sure that the string is parallel or close to parallel with the ground.

6. From this position, let go of the single washer. Be sure to keep your pointing index finger as still as possible. Wait . . . the washers didn't hit the ground. What happened? You'd better try it again to make sure that it wasn't a fluke.

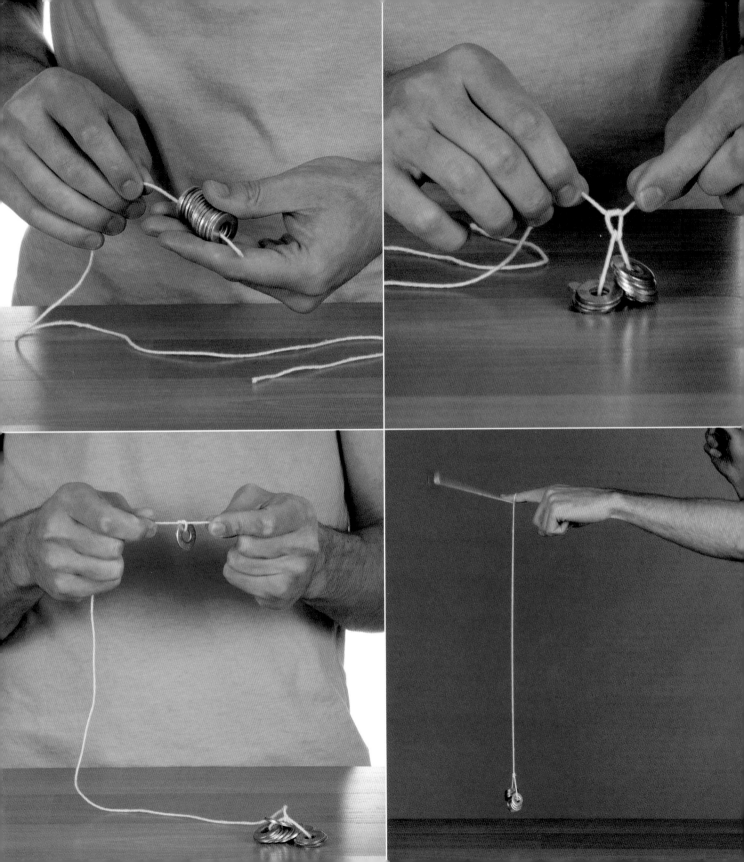

Try doing the experiment with one key tied to one end of the string and a key ring of 14 keys tied to the other end of the string. Drape the string over a pencil (instead of your finger) and see if the same thing happens. To make it easy, use keys that are identical. If you don't have 15 identical keys (and who does?), just make sure the keys are in approximately a 14-to-1 ratio (you'll learn more about this all-important ratio in a minute).

Here's another consideration. Try changing one variable at a time. Remember that you should change only one thing at a time so you can clearly see the effect of the change. If you change several variables at once, you won't know which one caused the change. What happens if you change the weight of the objects? The ratio of the objects? The kind of string? The length of the string? So many questions, so much science!

Experiment with changing the incline angle of the string before you drop the metal washers. When the string is at a much sharper angle does the same thing happen as when the string is horizontally parallel to the ground?

If you use a pencil instead of your finger, does the surface of the pencil make a difference in the result? What happens when a pencil has ridges instead of a smooth surface?

Try dropping the pendulum from different heights using different lengths of string between your finger and the washers. Is there a point, either too high or too low, at which the magic of your pendulum no longer works? Is there a height that works better than in our original experiment?

How many washers can you add to the lighter side of the pendulum and still have it work? How few can you have on the heavier side?

On a personal note, I was trying to explain this cool demonstration to a group of friends at a dinner party. Unfortunately, I didn't have a handful of metal washers, but we managed to scrounge up some string. Using only what we could find on the dinner table, the after-dinner scientists discovered the perfect relationship between a coffee cup and a spoon. This isn't to say that all coffee cups and spoons fall into the 14-to-1 weight ratio rule, but you get the idea.

We also discovered that a wedding band and a wine glass share a 14-to-1 relationship. Again, not all wedding bands and wine glasses will work, but it's fun to try! Tie the string to the stem of the wine glass and the wedding band. Drape the string over your finger, holding the wine glass upside down. Let go of the wedding band and watch what happens. Just be ready for some crash landings. If the coffee cup or the wine glass crashes into the table, try adjusting the weight on the other end of the string, the length of the string, or the incline angle of the string. This is yet another example of when you should follow our motto, **"Don't try this at home . . . try it at a friend's home."**

What if a television producer called you and wanted you to do this experiment on national TV? Dropping a washer isn't all that exciting. What if he wanted you to use a car on one side of the "string"? What might you have to use on the other side of the "string" so that the car didn't crash down onto an unsuspecting victim below? How long do you think the "string" would have to be and what should the "string" be made of? What would be the best incline angle for the "string"?

Granted, this is an extreme example, but if you're a teacher, you might consider posing the question to your class. You'd get some interesting answers, and it would become clear whether the students understood the importance of the 14-to-1 ratio and the concepts behind the Unbelievable Pendulum Catch activity.

The apparatus that you've constructed out of some string and metal washers is a **pendulum**. A pendulum is a weight suspended from a **pivot** (or fixed point) so that it can swing freely, back and forth. A common example of pendulums can be found in such timepieces as grandfather clocks. The pendulum is the long piece below the clock face that swings slowly back and forth.

Through a lot of trial and error, we've discovered that the magic ratio of weight between the two objects at either end of the string should be about 14 to 1. That's why you started with 15 metal washers. The key to making the trick more fun and interesting for your audience is to find two objects whose masses meet the 14-to-1 ratio requirement. Using the dinner table as the playground (so to speak), we found that our coffee cup and spoon obeyed the 14-to-1 rule. At this point, your wheels should be spinning as you think to yourself, "What object could be on the opposite end of the string from a 14-pound bowling ball?"

In terms of pure science, pendulums like the one you constructed operate using acceleration from gravity. When you release the single metal washer, gravity accelerates it toward the ground, giving it velocity. In a normal pendulum, the velocity decreases as the pendulum swings. The **amplitude** (how high the pendulum swings) also decreases the more the pendulum swings.

In our pendulum, the distance between the pivot (your finger) and the bob (the single washer) is decreased very rapidly when you release the string. As the distance between the bob and the pivot decreases, the velocity of the pendulum increases. With the velocity increasing so rapidly, its amplitude is increased to a point that it makes a number of full swings, wrapping the string around your finger. **Friction** keeps the group of metal washers from falling to the ground. With each turn around your finger, the friction increases and stops the fall of the large stack of washers.

Perhaps a simpler answer to why the dangling object seems to defy gravity is energy—kinetic energy and potential energy. **Kinetic energy** describes the energy of a moving object. **Potential energy** describes the stored energy in a stationary object. As in this activity, when you are holding the string tied to the metal washers in your fingers, both ends of the string have potential energy. Once you let go of the single washer, that potential energy turns into kinetic energy since the washer is moving. The washer falls a long way and builds up speed, or increases its kinetic energy, as it falls. The kinetic energy of the single washer is great enough to cause the end of the string to wrap itself around your finger.

MUST-SEE **SCIENCE**

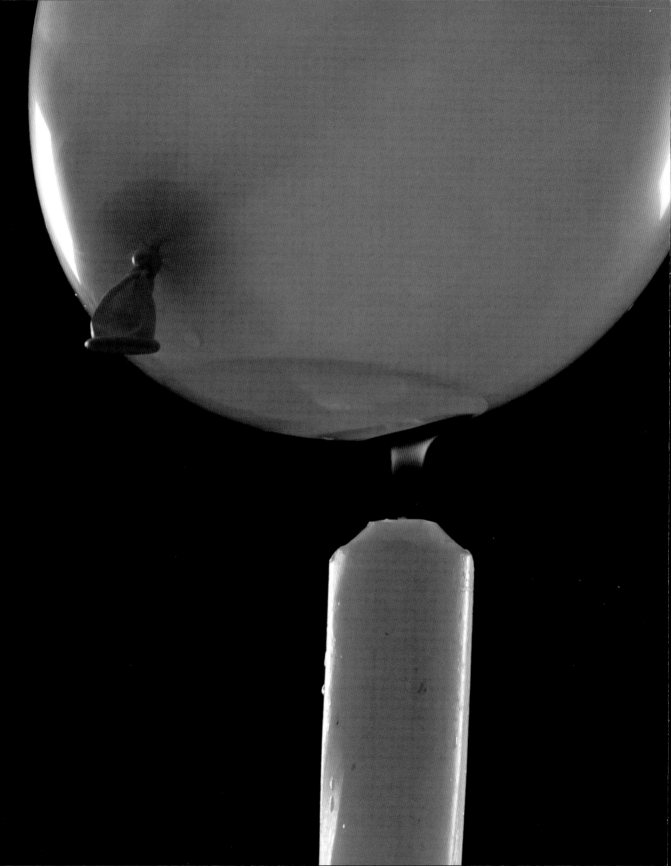

FIREPROOF **BALLOON**

Common sense tells you that it's impossible to boil water in a paper bag, but this classic parlor trick was a favorite of the Victorian Age magician. The real difficulty in performing this effect is making it look harder than it is! As you might imagine, the secret lies in yet another amazing property of water—its ability to conduct heat. Instead of using a paper bag, this modern-day version of the demonstration uses an ordinary balloon, some water, and a candle. It's a combination that's guaranteed to make people stand back. Besides, who doesn't love water balloons?

WHAT YOU NEED

Balloons

Water

Matches

Candle

Safety glasses

WARNING! This science activity uses matches, which means you need to find a very cool supervising adult to help with this experiment.

LET'S TRY IT!

1. Blow up a balloon just as you normally would and tie it off.

2. Light a candle and place it in the middle of the table.

3. Put on your safety glasses because it's time to pop the balloon. Hold the balloon a foot or two over the top of the flame and slowly move the balloon closer and closer to the flame until it pops. You'll notice that the flame doesn't have to even touch the balloon before the heat melts the latex and the balloon pops. Let's just say you had to prove what you already know.

Let's repeat the experiment, but this time the bottom of the balloon will have a layer of water inside.

1. Fill the balloon to the top with water—it probably holds a few ounces (that's 60 mL for you scientists out there)—and then blow it up with air. If you accidentally let go of the balloon before you tie it off, you'll spray yourself, and your friends will love it. Just tie off the balloon and get ready for the next step.

2. Hold the water-filled balloon at the top while you slowly lower it over the candle and watch as people start to run. Everyone knows that it's going to pop, but for some strange reason it doesn't. If you're very brave, you can actually allow the flame to touch the bottom of the balloon, but it still doesn't pop.

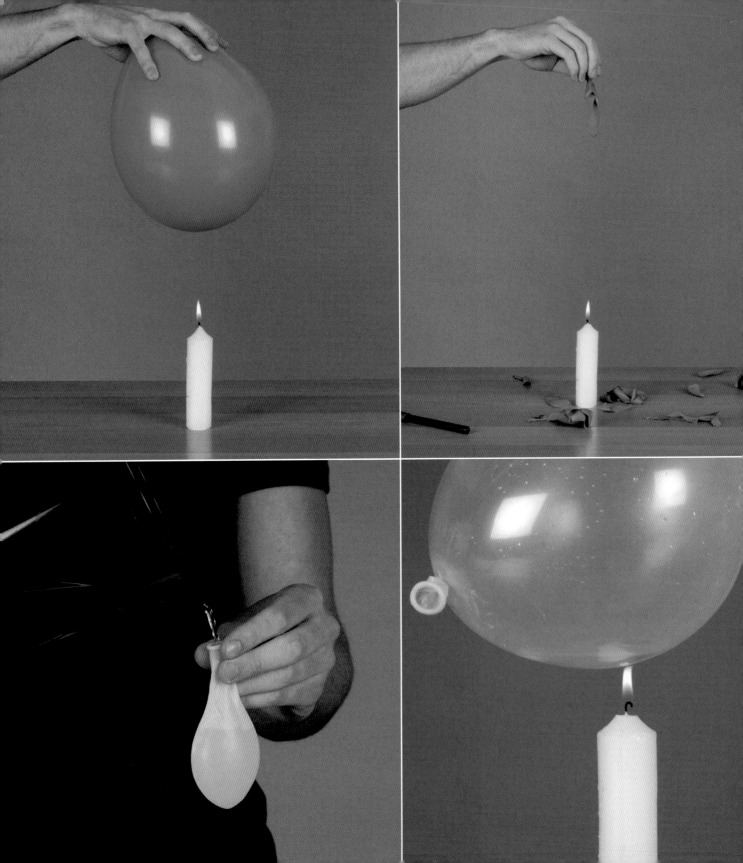

3. Remove the balloon from the heat and carefully examine the soot on the bottom. Yes, there's soot, yet the balloon didn't pop. Before reading the explanation, try to figure out why the layer of water kept the balloon from popping.

WHAT'S GOING ON HERE?

Water is a great substance for soaking up heat. The thin latex balloon allows the heat to pass through very quickly and warm the water. As the water closest to the flame heats up, it begins to rise and cooler water replaces it at the bottom of the balloon. This cooler water then soaks up more heat and the process repeats itself. In fact, the exchange of water happens so often that it keeps the balloon from popping . . . until the heat of the flame is greater than the water's ability to conduct heat away from the thin latex and the balloon pops. But watch out! If you turn the balloon so that the candle flame is close to the side of the water balloon, the balloon will pop because the water is not conducting the heat away from the surface of the balloon. At least the water will help put out the fire!

The soot on the bottom of the balloon is actually carbon. The carbon was deposited on the balloon by the flame, and the balloon itself remains undamaged.

REAL-WORLD APPLICATION

Using water to control heat is a valuable process. Special superabsorbent polymer foams are currently being used by firefighters as a way to help protect homes from being consumed by a raging forest fire. The water-absorbing polymer foam is similar to the superabsorbent polymer found in a baby diaper. The foam is applied like shaving cream to the outside of the house. As the fire burns closer and closer to the home, the water-filled foam absorbs heat energy from the fire and buys firefighters some extra time as they try to fight the flames with water.

Your body even uses water to control heat. When you exercise, your body produces sweat in an attempt to regulate your temperature so you don't get overheated. As the sweat evaporates, it takes heat energy with it, leaving cooler skin behind.

FEDERAL RESERVE NOTE

UNITED STATES OF AMER

S LEGAL TENDER
PUBLIC AND PRIVATE

F 9259

WASHING

63 A

SERIES
2006

ONE DOLLAR

IT PAYS **TO SMILE**

How often do you look at the face of the president pictured on your paper money? What's the facial expression of the president? Happy? Sad? Believe it or not, you can make George Washington's face smile or frown on a dollar bill if you know the secret to this visual foolery.

WHAT YOU NEED

A new, crisp $1 bill

LET'S TRY IT!

The secret is in the way that you fold the dollar bill. Just follow these simple steps.

1. Start with George's portrait facing you.

2. Make a "mountain fold" through the middle of George's left eye. In origami terms, this means to fold the bill away from you.

3. Make a second mountain fold through the middle of his right eye. Make sure that these creases are sharp.

4. Make a "valley fold" between the two previous folds so that the crease is between George's eyes and nose. To make a valley fold, fold the bill toward you. If all of these folds have you confused, check out the photos for clarification.

5. Pull on the ends of the bill slightly so that you can see his entire face, but make sure that the folds are still present.

6. Hold the portrait side of the bill in front of you with the face tilted upward. Notice how George smiles at you!

7. Slowly begin to tilt the bill downward, as if George were looking at the floor. Don't take your eyes off George's face because his smile will magically turn into a frown!

TAKE IT FURTHER

Use different dollar bills ($5, $10, $20, $50, even $100) to see if you can alter the emotions of some other presidents. Is Abe hiding a smile?

The George Washington facelift is a great example of an optical illusion. The opposing folds cause the features on the face to bend and contort, depending on how you look at them. Although it's one of the simplest activities in the book, it's the one that you'll find yourself doing over and over again.

During my years as an elementary school teacher, I had the privilege of serving as one of the student council sponsors. Anyone who has ever sponsored an extracurricular activity knows that it can be a real challenge to get kids to show up to school an hour early or stay late unless you have a "hook." In an attempt to get more kids to each meeting, I promised to start each meeting with a science demonstration. Things that fizz, bang, pop, or make kids oooh and ahhh were bound to get everyone to the meetings on time.

Little did I know that these gee-whiz demonstrations would do far more than just coerce kids to participate in student council. The other sponsor and I quickly learned the value of using the demonstration to illustrate a simple lesson in leadership or character building, and the smiling dollar bill illusion fit the bill (pun intended) perfectly.

After teaching the students how to make the George Washington on their own bill smile or frown, we asked them to get into small groups of three or four kids and come up with what leadership lesson they thought we were trying to illustrate using this demonstration. Truth be told, we were just fishing for their answers. Here are a few of their responses.

- Keep your chin up and share a smile. Cheer up!
- Lift your personality—smiles count!
- It makes good cents to smile.
- People can tell your attitude by looking at your body language.
- First impressions are often deceiving.
- Things are not always what they seem to be.
- It's hard to frown when you keep your chin up.
- Keep your eyes and nose pointed in the direction you want to go . . . UP!
- Wrinkles can give you a new perspective.

As you might imagine, the lunch line was a buzz of activity the next day as the student council kids taught their friends how to make George frown and smile. Let's just say the activity went "viral" before anyone knew what the term meant. However, the best part of the experience for us was discovering that our little leadership ambassadors were using the optical illusion to share their own leadership lessons with friends. What started out as just a clever trick to do with an ordinary dollar bill turned out to be an object lesson that truly left a lasting impression.

SKEWER THROUGH **BALLOON**

Some things in this world just don't mix—dogs and cats, oil and water, needles and balloons. Everyone knows that a balloon's worst fear is a sharp object . . . even a sharpened wooden cooking skewer. With a little scientific knowledge about polymers, you'll be able to perform a seemingly impossible task—pierce a balloon with a wooden skewer without popping it. Suddenly piercing takes on a whole new meaning!

WHAT YOU NEED

Several latex balloons
(9-inch size works well)

Bamboo cooking skewers
(approximately 10 inches long)

Cooking oil or dish soap

Sharpie pen

Nerves of steel

LET'S TRY IT!

1. The first step is to inflate the balloon until it's nearly full size and then let about one-third of the air out. Tie a knot in the end of the balloon.

2. If you carefully examine the balloon, you'll notice a thick area of rubber at both ends of it (where you tied the knot and at the opposite end). This is where you will pierce the balloon with the skewer, but not yet. Keep reading.

3. Coat the wooden skewer with a few drops of vegetable oil or dish soap (being careful not to accidentally get a splinter). As you probably guess, either liquid works well as a lubricant.

4. Place the sharpened tip of the skewer on the thick end of the balloon and push the skewer into the balloon. Be careful not to jab yourself or the balloon with the skewer. Just use gentle pressure (and maybe a little twisting motion) to puncture the balloon.

5. Push the skewer all the way through the balloon until the tip of the skewer touches the opposite end of the balloon, where you'll find the other thick portion of the balloon. Keep pushing until the skewer penetrates the rubber. Breathe a huge sigh of relief and take a bow!

6. Gently remove the skewer from the balloon. Of course, the air will leak out of the balloon, but the balloon won't pop.

Let's do it again, but this time you'll see the hidden "stress" in a balloon.

1. Before blowing up the balloon, use the Sharpie pen to draw about 7–10 dots on the balloon. The dots should be about the size of a dime. Be sure to draw them at both ends and in the middle of the balloon.

2. Inflate the balloon half full and tie the end. Observe the various sizes of the dots all over the balloon.

3. Judging from the size of the dots, where on the balloon are the latex molecules stretched out the most? Where are they stretched out the least?

4. Coat the skewer with a few drops of oil or dish soap to help ease the stick through the balloon.

5. Use the observations that you made previously about the dots on the balloon to decide the best spot to puncture the balloon with the skewer. Of course, the object is not to pop the balloon!

TAKE IT FURTHER

Pushing a skewer through the ends of the balloon is a challenging task, but attempting the same thing in the middle of the balloon is impossible . . . unless you have a few pieces of clear tape. Blow up a new balloon and place a small piece of clear tape in the middle of the balloon. Position the sharpened point of the skewer in the middle of the tape and carefully push the end of the skewer into the balloon without popping it! Try it again, but this time use a straight pin or the sharpened end of a safety pin. What role does the tape play in keeping the balloon from popping?

WHAT'S GOING ON HERE?

The secret is to use the portion of the balloon where the rubber molecules are under the least amount of stress or strain. If you could see the rubber that makes up a balloon on a microscopic level, you would see many long strands or chains of molecules. These long strands of molecules are called **polymers**, and the elasticity of these polymer chains causes rubber to stretch. Blowing up the balloon stretches these strands of polymer chains.

After drawing on the balloon with the Sharpie marker, you probably noticed that the dots on either end of the balloon were relatively small in comparison to the enlarged dots in the middle section of the balloon. You've just uncovered the area of least stress—the ends of the balloon. You wisely chose to pierce the balloon at a point where the polymer molecules were stretched out the least. The long strands of molecules stretched around the skewer and kept the air inside the balloon from

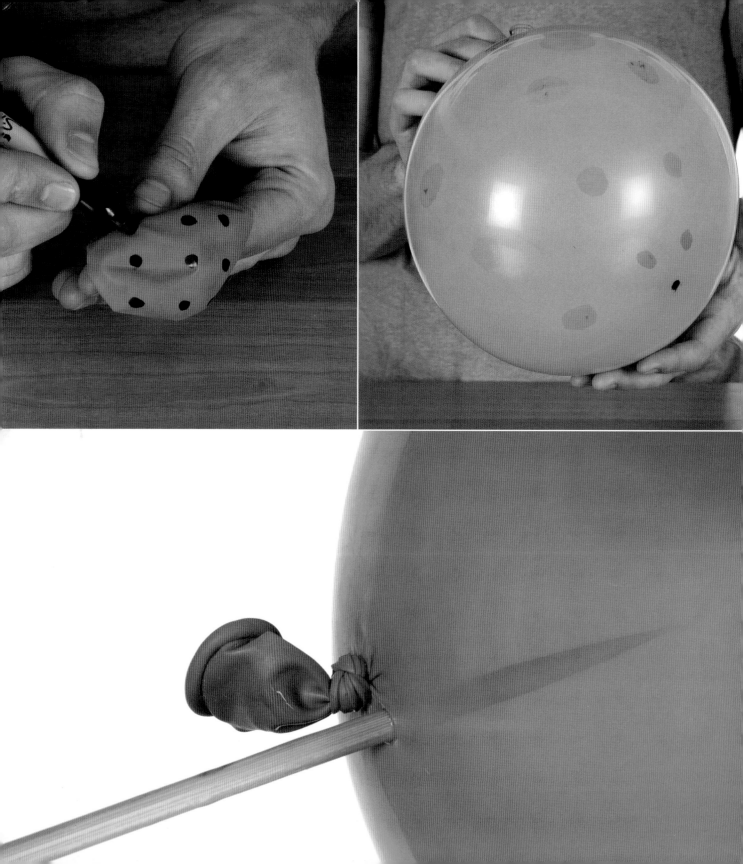

rushing out. When you removed the skewer, you felt the air leaking out through the holes where the polymer strands were pushed apart. Eventually the balloon deflated, but it never popped, right?

Trying to pierce the balloon in the middle section is nearly impossible unless you have some help from a piece of clear tape. Normally, the long chains of rubber molecules are under so much stress or tension that they tear easily with the slightest puncture. The clear tape helps to hold the rubber molecules in place, preventing them from tearing apart when the balloon is punctured. In an attempt to fool the people who know how to push a skewer through the ends of a balloon, magicians will secretly place pieces of clear tape in the middle section of a balloon to keep the balloon from popping when they puncture the balloon in this "forbidden" area. It's just another way to make even the smartest people say, "How did you do that?"

Aside from learning about the science of polymers, the Skewer Through the Balloon activity can be used as a great way to demonstrate a life lesson on how to approach a potentially stressful situation. The key to approaching any stressful situation is to find the area of least stress and to use this as an entry point as you attempt to relieve the tension. In other words, every stressful situation has a "best point of entry" and requires a good exit strategy if you are going to be successful in diffusing the potentially volatile situation. It's a great object lesson at any age!

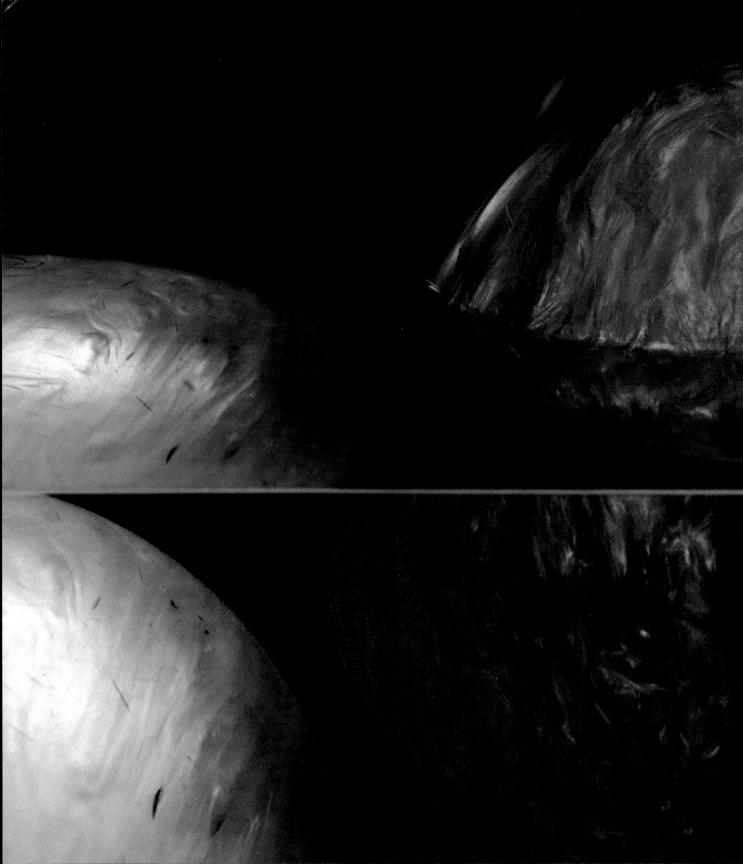

FLOATING BOWLING BALLS

We all know that certain things float in water while other things sink, but why? Do all heavy things sink? Why does a penny sink and an aircraft carrier float? Think you know the answers? Well, get ready for a few amazing surprises.

WHAT YOU NEED

The next time you're at the bowling alley, sneak in an aquarium filled with water under your coat. Too risky? Okay, just collect a few bowling balls of various weights (you need at least one bowling ball less than 12 pounds and one weighing 12 pounds or more). You might want to ask your local bowling alley if they are throwing away any old bowling balls that you might be able to use for your experiment.

A large aquarium, 5-gallon bucket, washtub, or bathtub filled with water

LET'S TRY IT!

1. Fill the aquarium three-fourths full with water.

2. Carefully place (do not drop) one bowling ball in the water. Does it float or sink? Repeat this experiment, noting the weight of each bowling ball, until every ball has taken the plunge.

3. What did you discover? It seems that anything heavier than 12 pounds sinks, but bowling balls less than 12 pounds float. Amazing!

TAKE IT FURTHER

Don't limit your curiosity to bowling balls alone. Try to float anything. Will a bottle of ketchup float or sink? Will a rubber chicken float or sink? How about an orange? Here's something strange. An unpeeled orange floats but a peeled orange sinks. Hmm, any guesses? Tiny pockets of air are trapped in the orange rind, making the unpeeled orange float in water. (These pockets of air work like "floaties" and decrease the density of the orange so that it floats in the water.)

This game is a great way for you to practice formulating a hypothesis, testing a theory, and using what you know (or don't yet know) about density to determine why an object floats or sinks, just like a real scientist.

Note: It's important that we work in metric units as we make the necessary density calculations to fully understand the science. *Pull out your calculators and sharpen up your math skills.*

If you've ever been bowling, you know that bowling balls range from about 8 to 16 pounds. However, did you know that all regulation bowling balls are the same size? According to official bowling rules from the Professional Bowlers Association,

> The circumference of a ball shall not be more than 27.002 inches nor less than 26.704 inches, nor shall it weigh more than 16 pounds (no minimum weight).

You could measure the circumference of each of your bowling balls by wrapping a piece of string around the ball at its widest part and marking the circumference on the string. Based on what we know from the Professional Bowlers Association, your measurement should be 27 inches, so that's the number we'll use for all of our calculations.

Since we're doing science, it's necessary to convert inches to centimeters using the conversion factor of 1 inch = 2.54 centimeters (cm).

So, the circumference of the bowling ball in centimeters is:

$$27 \text{ inches} \times 2.54 \text{ cm/inch} = 68.58 \text{ cm}$$

Next, we need to find the radius of the bowling ball. Again, you could use a string and a ruler to make your measurement, or you can use the formula for finding the circumference of a sphere to determine the radius. Remember that $\pi = 3.14$. Here's the formula:

$$\text{circumference} = 2 \times \pi \times \text{radius}$$

You know the circumference is 68.58 cm, so solving the equation for the radius you'll arrive at 10.92 cm.

Now that you know the radius, you can determine the volume using this equation:

$$\text{volume} = 4/3 \times \pi \times \text{radius}^3$$
$$\text{volume} = 4/3 \times 3.14 \times (10.92 \text{ cm})^3$$
$$\text{volume} = 5{,}451.75 \text{ cubic centimeters (cm}^3)$$

Of course, the keen observer will never fail to point out the fact that there are holes in the bowling ball for your fingers. Again, you could jump through some hoops to determine the volume of the holes, but it's fair to say the holes do not affect the overall density of the bowling balls enough to change the demonstration. So, we'll go with the magic number of 5,452 cm³ (yes, we rounded up to simplify the calculations).

As soon as you know the volume of the bowling ball, you can easily calculate the ball's density by dividing the weight of the ball (in grams) by the volume (5,452 cm³).

To convert the weight of the ball from pounds to grams, use the conversion of 1 pound = 453.6 grams (but we'll round up to 454 grams). Here's a table with the converted weights of bowling balls.

BOWLING BALL WEIGHT IN POUNDS	CONVERTED TO GRAMS
8 pounds	3,632 grams
9 pounds	4,086 grams
10 pounds	4,540 grams
11 pounds	4,994 grams
12 pounds	5,448 grams
13 pounds	5,902 grams
14 pounds	6,356 grams
15 pounds	6,810 grams
16 pounds	7,264 grams

Now it's easy to calculate the density of each of these bowling balls. Simply divide the weight of the ball in grams by the volume of the ball in cubic centimeters. For example, here's how to calculate the density of an 8-pound (3,632 g) bowling ball.

density = mass / volume
density = 3,632 grams / 5,452 cm³ = 0.67 g/cm³

It's no wonder that an 8-pound bowling ball easily floats in water. The density of water is 1.0 g/cm³ so the 8-pound bowling ball floats because it is less dense than the water (.67 < 1.0).

The chart below shows the calculated density of each size bowling ball, and you can see at what weight the bowling balls begin to sink (their density is greater than that of water at 1.0 g/cm³).

BOWLING BALL WEIGHT IN POUNDS	CONVERTED FROM POUNDS TO GRAMS	DENSITY WEIGHT / 5,452 CM³
8 pounds	3,632 grams	0.67 g/cm³
9 pounds	4,086 grams	0.75 g/cm³
10 pounds	4,540 grams	0.83 g/cm³
11 pounds	4,994 grams	0.92 g/cm³
12 pounds	5,448 grams	0.99 g/cm³
13 pounds	5,902 grams	1.08 g/cm³
14 pounds	6,356 grams	1.17 g/cm³
15 pounds	6,810 grams	1.25 g/cm³
16 pounds	7,264 grams	1.33 g/cm³

If the calculation is correct, a 12-pound bowling ball should just barely float in the water-filled aquarium.

If I lost you somewhere in the midst of all of these numbers, here's the synopsis. Any bowling ball that weighs more than 12 pounds will sink in water, and any bowling ball that weighs less than 12 pounds will float. If that wasn't enough, there's something else to consider. **Archimedes' Principle** states that the buoyant force exerted on a fluid is equal to the weight of the fluid displaced. So, when an object is placed in water, it will displace its weight in water. The 8-pound ball displaces 8 pounds of water. However, the ball also takes up more volume than 8 pounds of water, so it floats.

The same principle explains why an aircraft carrier floats even though it's made of steel. If you consider that steel has a density of 7.8 g/cm³, shouldn't the aircraft carrier sink? The ability of the ship to float is not entirely based on the material from which it is made. A ship is built in such a way that it encloses large amounts of open space, and this contributes to its overall density. According to Archimedes, the ship displaces its weight in water, but because of the way the ship is constructed, it takes up more space than the volume of the water it displaces and it floats.

THE SUPER-SECRET
TEACHERS-ONLY SECTION

DEMONSTRATIONS GUARANTEED
TO GET OOOHS & AHHHS

With a book title like *Fire Bubbles and Exploding Toothpaste*, these pages—as you've probably guessed—do not all describe experiments for young scientists to try at home. The truth of the matter is that young scientists just don't sprout up out of the ground each fall waiting to be harvested by the teachers at the local school. These little ankle-biters need to have someone spark their imagination, make them wonder, ask questions, pique their curiosity, and whet their appetite for wanting to learn more. In other words, kids need great teachers to help them grow into budding truth-seekers who call themselves future scientists. For this very reason, we've dedicated the last part of the book to teachers only.

If you are an adult reading this book, I hope you had the privilege of learning from a great chemistry or physics teacher at some point on your march to graduation. You probably don't remember the periodic table or physics equations, but if you had a great teacher, I'm sure you have memories of the day the flask exploded, the desk caught on fire, or the liquid nitrogen spilled everywhere. It wasn't that your teacher was irresponsible or unprofessional; instead, it was that your teacher was actively engaged in creating unforgettable learning experiences for his or her students. At a recent high school reunion (I won't tell you which one), those memories were the subject we discussed the most: "Do you remember when Mr. Hodous . . ." or "Wasn't it hilarious when Mr. Hogan . . ." Believe me, there certainly wasn't any discussion of how much we remember about *The Grapes of Wrath* or *Lady Chatterley's Lover* (no offense to English teachers, John Steinbeck, or Lady Chatterley).

If you read nothing else in this section, read this!

This "teacher-only" section is for certified, hands-on professionals . . . not teacher wannabes, parents, or kids. The activities are more intensive, require a higher level of thinking when it comes to doing science, and sometimes require having access to twisty-turny things, fire extinguishers, methane gas, safety shields, smelly solvents, beakers, flasks, and an overwhelming passion for wanting to make a kid's head explode with excitement when she screams the words, "Science is so cool!"

Teachers, this is the stuff memories are made of, so go ahead and try some of these activities to turn some gee-whiz demonstrations into true learning experiences for your students. You'll be the one everyone is talking about at the next reunion!

> Science teachers have a unique profession—one that uses demonstrations to build unforgettable experiences and to get kids to think. Great teachers are near and dear to all of us who had a part in making this book, so we wanted to include a section just for you.

—Steve Spangler

P.S. So, let's just say that for some reason you're a kid, and without hesitation, you immediately flipped ahead and glanced at such secret experiment titles as *Fire Tornadoes* or *Exploding Toothpaste*. You've just been caught red-handed, adding insult to injury. Now what? As your punishment, you are ordered to give this book to your favorite science teacher with bookmarks on the page of the experiments you want him or her to do in class. Please go out and purchase another copy of this book for yourself and we'll forget this ever happened. Deal?

FIRE BUBBLES

When is the last time you made bubbles catch on fire? Probably never . . . but then again, this is the *teachers-only* section of the book, so anything is possible. Fire Bubbles is the signature activity for this book (because it appears on the cover!), but it's also one of my personal favorites and one that I perform on television all the time. I honestly never thought that talk show host Ellen DeGeneres would allow me to light her hands on fire, but she did . . . and the audience went wild! Aside from the gee-whiz factor of Fire Bubbles, this demonstration brilliantly illustrates the heat-conducting properties of water and shows how even a very thin layer of water protects your skin from the blistering flames.

WHAT YOU NEED

Access to methane gas

Plastic jar

Dish soap

Rubber tubing (3 feet long)

Beach ball

Large bowl or a plastic storage container half filled with water

Lighter (Something like a striker or an aim-and-flame works great.)

Two pairs of safety glasses (one for you and one for your helper)

Fire extinguisher

Someone to help you

LET'S TRY IT!

1. You'll want to take the beach ball with you to the hardware store as you look for the right size of rubber tubing (you'll find it in the plumbing section). Your mission is to select tubing that will tightly fit into the opening of the beach ball. It should be a very snug fit to keep the gas from leaking out. A helpful tip is to put a few drops of dish soap on the end of the tubing to act as a lubricant as you push the end of the tubing into the beach ball. Just feed the tube in a few inches in order to secure it in place.

2. It's best to start by finding a clear plastic jar similar to the one pictured in the photos. Start by filling the plastic jar about one-half full with ordinary tap water. Add a squirt of dish soap to the water.

**WARNING!
TEACHERS ONLY!**
This demonstration is provided for educational purposes only and should not be attempted if you are not a properly trained professional. If you choose to perform this demonstration, safety glasses, a fire extinguisher, and a friend to help you are all required.

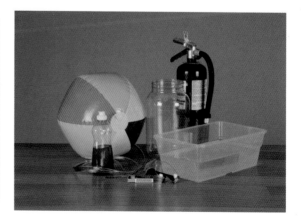

3. It's important to make sure that the beach ball is empty (all of the air is squeezed out) before filling it with methane gas. Again, you'll need to have access to a natural gas jet (the gas is methane) in the school chemistry lab. Hold the tube up to the jet and fill the beach ball with methane gas. Place your finger over the opening of the tube (or use a bulldog clip to crimp the hose) to keep the gas from escaping while you transport it over to the jar of soapy water.

4. Submerge the open end of the tube in the soapy water. Secure the tube at the opening of the jar with a piece of duct tape, or you can simply drill a hole in the plastic jar beforehand and feed the tubing through the opening as a way to keep it in place. The latter method is shown in the accompanying photos. Don't worry about a small amount of the gas coming out of the beach ball when you submerge the tubing. It's normal for this to happen.

5. It's time for your friend to step up and help out. Both of you need to put on your safety glasses. Your helper is in charge of squeezing the beach ball in order to generate bubbles of methane gas in the plastic jar. Squeeze the ball with gentle but constant pressure to generate some bubbles.

6. Fill the large bowl or plastic storage container one-half full with plain water.

7. While your friend is squeezing the beach ball, submerge both of your hands in the container of plain water. The goal is to make sure every part of your hands and wrists are covered with water. Failure to get your hands and wrists completely wet can result in a burn . . . so make sure to douse them with water.

8. By this time, the bubbles of methane gas should be rising out of the plastic jar. Scoop up a small handful of bubbles with your left hand. Step away from your friend. Light the striker with your right hand and touch the bubbles in your left. (If you choose to scoop up bubbles in both hands, then your beach ball friend will have to be the one who lights the fire.) This is probably the scariest step in this entire book of demonstrations! The methane-filled bubbles will produce a large flame (keep your hand in front of you and don't move!), but it will extinguish on its own in just a few seconds, and your hands will remain unharmed! Scream with excitement and take a well-deserved bow.

9. Within seconds, you'll hear the chant, "Do it again, do it again!" Give your fans what they're asking for and do it again, but first make them offer up their best hypothesis as to why you didn't get burned. And make sure you douse your hands in the container of plain water again before your encore performance.

If you've attempted the Fireproof Balloon activity, you already have a good idea as to how this works. If not, it's a great idea to present the Fireproof Balloon activity prior to performing Fire Bubbles to see if any of this science is actually sinking into the brains of your audience members. Water is a great conductor of heat, and it literally pulls the heat away from your hand for the few seconds that the methane bubbles are on fire. As the flame heats the water on your hands, some of the water at the very surface evaporates, and it's this natural cooling process from the evaporation that keeps your hand from burning.

It must be said that if you play with fire, you'll eventually get burned. I wish that I could call it **Spangler's Law of Doing Science Demonstrations That Use Fire**, but I think that it's probably attributed to someone else. Nevertheless, you must use extreme caution when doing this activity. It's only natural to grow more and more confident as you repeat the demonstration, and you're likely to want to set fire to a huge column of bubbles on your hand. **Don't do it!** If the heat from the flame is greater than the water's ability to pull the heat energy away from your hand, you'll get burned (literally and figuratively).

When presented in the appropriate learning environment, using all of the necessary safety equipment (and common sense), this demonstration is guaranteed to become one of your all-time favorites.

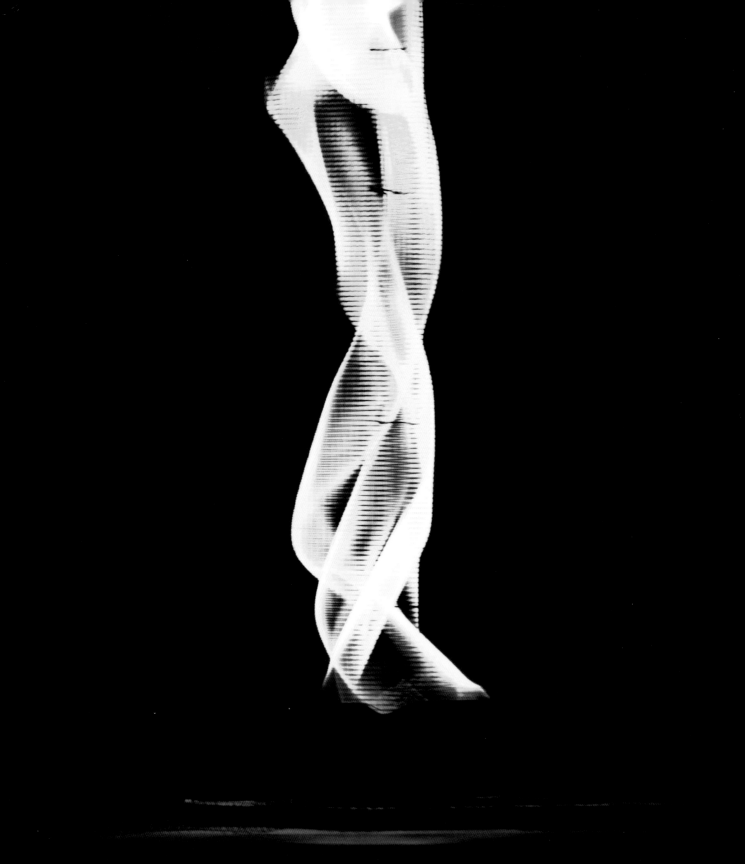

FIRE **TORNADO**

When we picture a tornado, most of us imagine a whirling column of air poking down from the clouds. But this tornado-like effect is not just limited to air. Imagine what it might look like if winds could twist a ground level forest fire into an enormous fire tornado that dances across the tops of the trees. It's not a special effect found in a movie—it's a real-world danger that firefighters battle in the most extreme forest fires imaginable. This special demonstration is reserved for science teachers who want to share a small version of this amazing phenomenon.

WHAT YOU NEED

Lazy Susan (rotating tray)

Metal screen (You'll want a mesh size that is similar to the screen found on your windows. The actual size of the piece of screen will depend on the size of the rotating tray.)

Wire or staples

Small glass dish

Glass dinner plate or something similar to cover the fire

Pieces of sponge

Lighter fluid

Safety glasses

Fire extinguisher

LET'S TRY IT!

1. The key to making this demonstration work well is to test a number of circular rotating trays (commonly referred to as Lazy Susans) until you find one that suits you. It's truly a matter of trial and error, but you're bound to find one that functions perfectly.

2. Take a look at the photographs of the wire screen that is rolled into the shape of a tube. Find a friend to help you shape and hold the wire in place as you roll the screen into a cylinder that's about the same diameter as the rotating tray. When you're constructing this tabletop version of the Fire Tornado, it's best to keep the height of the screen tube between 2.5 and 3 feet tall. Anything taller has the potential of falling over as you spin it. As you can see in the photographs, our wire cylinder rests on top of the tray and is just slightly smaller than the tray itself. Fasten the ends of the cylinder using wire or staples. Rivets or wire can be used to secure the center section.

3. Position the wire cylinder in the middle of the tray and give it a gentle spin. You should be able to make the cylinder spin slowly without having to physically fasten the screen to the tray. When you're presenting the demonstration, you'll need to be able to quickly remove the screen from the rotating tray and cover the fire with a plate to extinguish the flames.

4. Place the small glass dish in the middle of the rotating tray. It is best to find a small square of fire-resistant material (or a small plate or saucer) for the dish to sit on so as not to damage the Lazy Susan.

5. Cut up several pieces of sponge and place them in the dish. Squirt lighter fluid on the pieces of sponge so that each piece is completely soaked.

6. Put on your safety glasses. Light the fire but leave the mesh screen off of the rotating tray for now and gently spin the tray. Notice how the fire spins, but no tornado effect is created. Extinguish the fire by covering it with another small plate.

7. Reignite the fire and place the wire screen cylinder on the Lazy Susan. Gently spin the tray and watch as the fire twists into the shape of a tornado. The fire tornado will rise as the tray spins faster and faster.

8. Remove the screen cylinder from the tray and extinguish the fire.

Experiment by making wire cylinders that are made from different types of metal screen—large mesh or small mesh constructed out of thin or thick wire. Each variation will produce a different effect. Don't be surprised if the thick wire, large mesh screen doesn't work at all. Why?

Visit your local thrift store, garage sale, or flea market in search of an old record player. With a little modification, you'll be able to transform the turntable into the spinning platform needed for your experiment.

WHAT'S GOING ON HERE?

As you noticed, simply spinning the tray does not whip the fire into a twirling tornado. It's only when you center the fire in the middle of the rotating screen that you create the perfect fire tornado. It all starts with the heat from the flame that causes the surrounding molecules of air to rise. Couple this with the rotational motion of the screen and you have the perfect storm, so to speak. The rotating screen gives the air molecules an initial spin (called **angular momentum**). The vertically rising hot air molecules collide with the rotating screen, and the angular momentum of the screen is transferred to the rapidly rising air molecules, giving them a "twist." Fresh air fuels the fire from the bottom, and the flames twist into the shape of a tornado.

REAL-WORLD APPLICATION

The rotating metal screen is a simple way to illustrate the way winds whip through the trees in the forest and collide with the warm updraft from the wildfire. These so-called fire tornadoes can measure 30–50 meters tall (100–200 feet). Some of the largest fire tornadoes have measured more than a kilometer in height. It's easy to see how the rising column of twisting fire can dance along the treetops, causing the fire to spread easily.

VANISHING **PEANUTS**

You won't believe your eyes when you see what happens to ordinary Styrofoam packing peanuts when they come in contact with a solvent called acetone. They seem to magically "disappear." In fact, the Styrofoam reacts with the solvent to reveal the fact that Styrofoam is made up of long strands of styrene molecules with lots of air pockets. This demonstration also reminds us about the importance of reducing our use of Styrofoam and replacing it with more Earth-friendly packing materials.

WHAT YOU NEED

Styrofoam packing peanuts

Starch-based packing peanuts

2 bowls

Styrofoam cups

Glass jar

Acetone solvent (available at paint stores and for adults ONLY!)

Styrofoam head used to display wigs

A 3/4-inch thick sheet of Styrofoam insulation (found at your local hardware store in the area where insulation is sold)

250-milliliter beaker (or an 8-oz glass)

Knife

Drill

Sharpie marker

Safety glasses

LET'S TRY IT!

1. Fill two bowls one-half full with water.

2. Put a handful of Styrofoam peanuts in one bowl and a handful of starch-based peanuts in the other bowl.

3. Notice how the Styrofoam does not dissolve.

How to Make Styrofoam Packing Peanuts Vanish

1. Fill a beaker with 60 milliliters (that's about 2 oz) of acetone.

2. Drop Styrofoam peanuts into the beaker and watch what happens. It looks like they're melting, but they're really just dissolving (melting requires heat). You'll be amazed to see how much Styrofoam will dissolve in just a few ounces of acetone.

3. If you have a large enough beaker, try placing a Styrofoam cup into the beaker to see what happens. With just a little swirl, the entire cup will vanish in seconds.

Air Head—Vanishing Styrofoam Strips

1. Place a Styrofoam head used to display wigs on the table. The next step is the trickiest one—to carve a huge hole in the top of the Styrofoam head. You want the hole big enough to hold the 250-milliliter beaker. For those nonmetric people, you'll want a glass that holds about 8 ounces of liquid. You can use an electric drill with a doorknob hole cutter blade to get the hole started, but it's going to take a little patience until the hole is just the right size.

2. Fill the beaker with 200 milliliters of acetone (about 6 oz) and carefully lower the beaker into the hole. Be careful not to spill any acetone on the Styrofoam head or it too will dissolve! You might also want to put safety glasses on the Styrofoam head, just for good measure.

3. You'll need a sharp knife to cut the Styrofoam board into long strips. The width of each strip is determined by the diameter of the glass container in the head (250-mL beaker or otherwise). Cut as many strips as you feel the urge to make disappear. *NOTE: Some Styrofoam board material has a thin, plastic covering on both sides. Remove any plastic wrapping before doing the demo.*

4. Use a Sharpie pen to write down anything important you want the "head" to know. Consider writing the three R's of recycling on the three strips—REDUCE, REUSE, and RECYCLE. This would be a great activity to do as a review for a test to "cram" the most important information into the Styrofoam head *and* your students' heads.

5. It's showtime! The story line is up to you, so be creative. When it's time to make the strip vanish, slowly push the strip into the beaker of acetone, being careful not to make the acetone erupt onto the Styrofoam head. The illusion is great, as it looks like knowledge is being shoved piece by piece into the head.

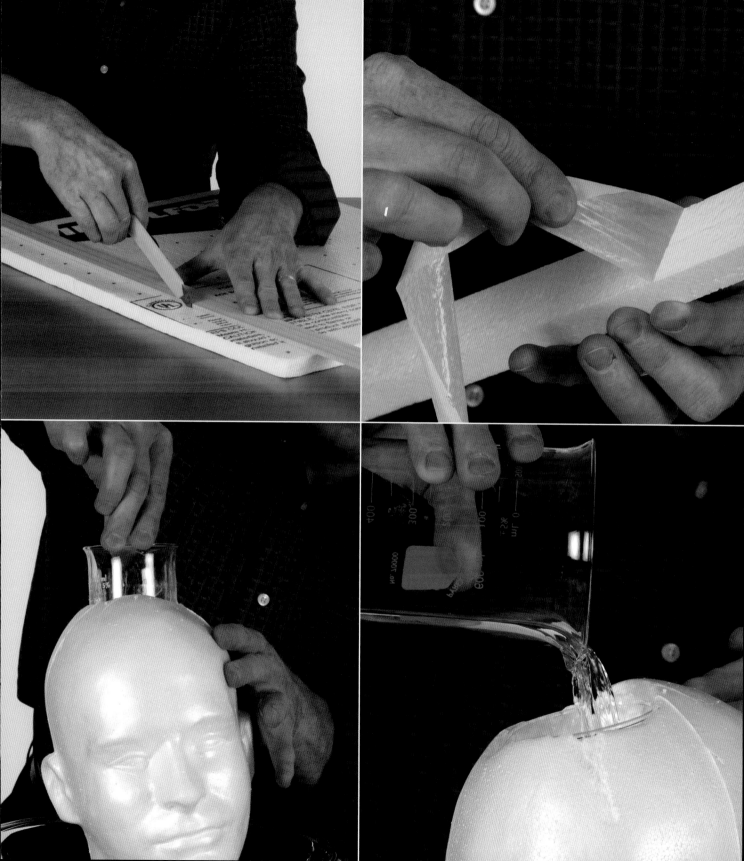

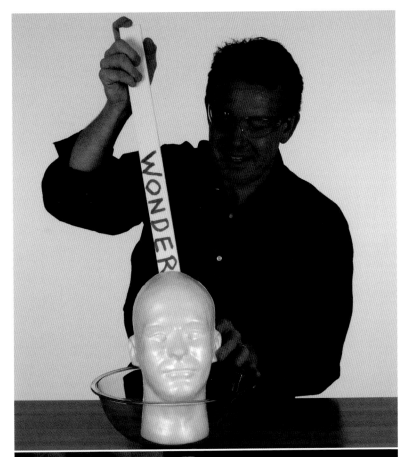

Here's a fun activity for teachers. Engage students in a peanut race by seeing which team can fill a bucket first with polystyrene peanuts. Of course, one bucket will secretly contain acetone and the team with this bucket hasn't got a prayer of winning! Use extreme care when handling acetone. Follow the manufacturer's directions for proper use and disposal.

Colored Starch Noodles

Several companies have turned starch peanuts into an arts and crafts project for kids by simply adding color to starch peanuts and calling them "building noodles." Here's how it works: simply wet one end of the colored starch peanut with a dab of water and stick it to another peanut. Build houses, hats, glasses, letters, people, a medieval castle with flying buttresses . . . just build anything! Use the colored starch peanuts as an icebreaker or team-building activity with adults or kids. They're great for staff meetings!

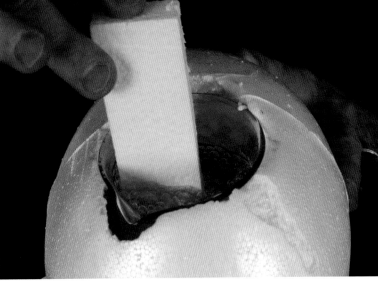

Styrofoam is actually a common name for a material called polystyrene, which is a polymer made up of a long chain of molecules. The packing "peanuts" are made in a special process whereby air is blown into liquid polystyrene to make puffy peanut-shaped pieces out of a very tiny quantity of polystyrene. If you break apart a Styrofoam peanut, you'll see little pockets of air that serve as a cushion to protect delicate items that are packed and shipped. The acetone is a solvent that easily breaks down the polystyrene, releasing the little air pockets and leaving very little residue at the end. In other words, the polystyrene dissolves in the acetone.

As a science teacher, science enthusiast, or an environmentalist, you are aware of the bad effects that Styrofoam has on our environment. That's why many companies have turned to starch packing peanuts as a substitute for Styrofoam. Instead of taking up space in the landfills, starch peanuts dissolve in water to make landfill gravy! So why not just use acetone to dissolve waste polystyrene? Problem solved, right? Not quite, since the acetone presents its own environmental and energy consumption issues.

REAL-WORLD APPLICATION

You Have a Choice

Currently about 200 million cubic feet per year of polystyrene "loose fill" (packaging material) is used in the United States. Although some companies try to reuse the packing material, most of the polystyrene loose fill is disposed of in a landfill. As students of science, we need to carefully examine such products and ask these questions: How is the material made, and what happens to it after it is used? One of the properties of polystyrene loose fill is that it does not compress easily. While this is beneficial when trying to protect something from being crushed or broken, it poses a problem when trying to dispose of it in a landfill. As a result, environmentally conscious companies sought a solution to these problems. One such solution is called Eco-Foam loose fill. It provides the ease of use and cushioning of polystyrene, but it gives many other reuse or disposal options for the future. It readily decomposes in water and can be reused for your own packages, or you can dispose of it by putting it in your compost pile, watering it into your lawn, or washing it down the sink.

Eco-Foam is made almost entirely from an annually renewable resource . . . corn! The remaining ingredient is a water-soluble organic polymer called "polyvinyl alcohol." This organic polymer is made from carbon, hydrogen, and oxygen—the building blocks of life. When polyvinyl alcohol is exposed to water, naturally occurring bacteria feed on this organic polymer. Under wet conditions, the bacteria will use the starch (which is also composed of carbon, hydrogen, and oxygen) and polyvinyl alcohol as food to begin the cycle of life again.

Many people feel that the answer to our solid waste problem is recycling. While this method will go a long way to help our solid waste problems, it is not the whole solution. One good suggestion is to use as little of the material as possible. Secondly, it makes sense to use a natural product (instead of a synthetic product) that will break down when we are finished using it. We must remember how to reuse!

EXPLODING TOOTHPASTE

Mix two solutions together and you get an amazing eruption of foam that looks like a giant stream of toothpaste exploding from the cylinder. Some people refer to this foam as Elephant's Toothpaste (when the reaction is in action, this name will totally make sense). We call it Exploding Toothpaste. Regardless of what you call it, this classic reaction is a favorite of chemistry teachers who have access to these chemicals that you will not find around the house. We've also included a kid-friendly version of Exploding Toothpaste that uses easy-to-find materials. This demonstration is guaranteed to produce a room full of ooohs and ahhhs the moment the foam begins to erupt from the bottle.

WHAT YOU NEED

Hydrogen peroxide (30%)

Sodium iodide crystals
(This is a dry chemical
that looks like salt.)

250-milliliter beaker

Liquid dish soap

Food coloring

1,000-milliliter
graduated cylinder

Measuring spoons

Safety glasses

Plastic tarp to cover the
demonstration table

Rubber gloves

LET'S TRY IT!

1. The first step is to put on your safety glasses. Since 30% hydrogen peroxide will burn if it comes in contact with your skin, it's best to wear rubber gloves to protect your hands.

2. Fill the beaker with 4 ounces (that's approximately 120 mL) of room temperature water. Add about a tablespoon of sodium iodide crystals to the water and stir with a spoon until all of the crystals have dissolved. Repeat this several times until the crystals no longer dissolve in the water. When this happens, you have what is called a **saturated solution**. Label the beaker "Sodium Iodide Catalyst" and set it aside to use later.

3. Cover the table with the plastic tarp to make cleanup easy at the end of the demonstration.

4. Measure 2 ounces (that's 60 mL) of the 30% hydrogen peroxide into the graduated cylinder. Position the graduated cylinder in the middle of the plastic tarp.

5. Add a squirt (that's a *very* technical term meaning about 5 milliliters) of dish soap to the graduated cylinder containing the 30% hydrogen peroxide.

6. Add a huge squirt of your favorite food coloring to spice things up. Give the solution a quick swirl to mix the contents.

7. The last step is to pour a tablespoon (that's about 5 milliliters) of the sodium iodide catalyst into the graduated cylinder and to quickly stand back. Within seconds, the reaction will occur and a mountain of erupting foam will cover the table. *Note: You can even take this basic reaction one step further by switching out the graduated cylinder for a large Erlenmeyer flask, creating an even bigger and better reaction.*

Everyone will want to touch the foam on the table, but you must keep the eager ankle-biters away just in case some of the hydrogen peroxide did not react with the catalyst. You don't want anyone to get his or her hands burned or stained by touching the foam.

All of the aftermath from this reaction is safe to either throw away in the trash can or wash down the drain.

How to Get the Stains Out of Carpet

If you perform this demonstration enough times, you'll have a situation where the foam spills over onto the floor. Unfortunately, the iodine that is released as part of the reaction will stain the floor or carpet, unless you know something about removing iodine stains. The secret is OxiClean®. This oxidizing, stain-removing detergent is available at any grocery store, and it does wonders for removing iodine stains. As the pitchmen say, "Use the power of oxygen to remove stains in an instant!" . . . and it works.

TAKE IT FURTHER

Concentration Variations

Because it is often hard for anyone but a chemistry teacher to obtain 30% hydrogen peroxide, you might want to test the effectiveness of other concentrations of hydrogen peroxide. For instance, 3% hydrogen peroxide is the household concentration that would typically be used for cuts and scrapes. It is safe to touch, but powerful enough to kill bacteria, viruses, and fungi on surfaces. Hair stylists can purchase peroxide that is anywhere from 6% to 12% in strength.

To make four different solutions, start with 30% hydrogen peroxide and dilute it by one-half to make a 15% solution. Dilute a small portion of the 15% solution by one-half to arrive at a 7.5% solution. One more dilution will produce a solution that is approximately 3.5%. Now you have the four solutions needed for the experiment. It is possible to purchase various concentrations of hydrogen peroxide at drugstores,

beauty supply stores, and chemical supply stores, but this dilution method is easy to do if you start with a good supply of 30% hydrogen peroxide.

The most important part of the experiment is that everything stays the same except for the concentrations of hydrogen peroxide. This simple process is a very effective way to illustrate how a scientist would control a single variable (the strength of the hydrogen peroxide) to affect the outcome of the reaction. Each cylinder should have 60 milliliters of hydrogen peroxide, an equal sized squirt of dish soap, and an equal amount of the sodium iodide catalyst. The only variable you want to change is the concentration of hydrogen peroxide; otherwise, there is no way to know which variable causes the effect in the experiment. Remember, a good science experiment changes only one variable at a time.

If you don't have four graduated cylinders, you'll need to set up the experiment four times and collect the data after each eruption. If you do have four graduated cylinders, it's easy to observe the differences in each eruption if you add the sodium iodide solution to each cylinder one right after the next. How does the concentration of the hydrogen peroxide affect the amount of foam that is created in each reaction?

A Halloween Twist on the Exploding Toothpaste Experiment

The classic Exploding Toothpaste experiment takes a whole new twist when you see it oozing from the face of your jack-o'-lantern!

This is what happens when chemistry teachers get tired of doing the same old Exploding Toothpaste demonstration over and over again. Instead of using a graduated cylinder, simply use a glass beaker to hold the 60 milliliters of hydrogen peroxide along with the squirt of dish soap and the food coloring. Place the beaker down inside a carved-out pumpkin and you're almost ready to go. The only thing left to do is to make sure your safety glasses are on and the kids in the front row have moved back to the third row. After adding the tablespoon of saturated sodium iodide solution, immediately replace the lid of the jack-o'-lantern and wait for the kids to scream. The foam will ooze from the eyes, nose, and mouth of the pumpkin, and you'll come away with a new discovery . . . ooze = ooohs!

WHAT'S GOING ON HERE?

You might remember Mom treating your scraped knee or a cut with hydrogen peroxide. H_2O_2 is the scientific name for hydrogen peroxide, which is made up of two hydrogen atoms and two oxygen atoms. H_2O_2 looks like ordinary water (H_2O), but the addition of that extra oxygen atom turns the molecule into an extremely powerful oxidizer. The hydrogen peroxide used in this demonstration is ten times stronger than the over-the-counter hydrogen peroxide you can find at the store. Low-grade hydrogen peroxide (3%) will not produce the massive amount of foam seen in this version of the Exploding Toothpaste demonstration. The secret ingredient is actually sodium iodide, which acts as a **catalyst** (something that speeds up a chemical reaction, and in this case, it's the decomposition of hydrogen peroxide). When hydrogen peroxide (H_2O_2) decomposes, it breaks down to form water (H_2O) and oxygen (O_2). The soap bubbles that erupt from the cylinder are actually filled with oxygen. You'll notice that the foam has a brown tint. This color is due to the presence of free iodine produced by the extreme oxidizing power of the 30% hydrogen peroxide. As the reaction takes place, you'll also see steam rising from the erupting foam. This shows that the reaction is **exothermic**, meaning that it gives off heat.

Hydrogen peroxide (30% strength) will act as an oxidizing agent with practically any substance. This substance is severely corrosive to the skin, eyes, and respiratory tract. Sodium iodide is slightly toxic by ingestion. Given these safety precautions, it's best to leave this one to the experts. Just befriend a chemistry teacher and ask her to perform the famous Exploding Toothpaste experiment.

KID-FRIENDLY
EXPLODING TOOTHPASTE

This is a kid-safe version of the popular Exploding Toothpaste demonstration using materials that are easier to find. A child with a great adult helper can perform this activity safely, and the results are wonderful.

WHAT YOU NEED

1-liter plastic soda bottle

Hydrogen peroxide (12%) (This is found at a store that sells hair care products. Ask for hydrogen peroxide that is labeled 40-volume. This is the same as a 12% solution.)

Liquid dish soap

Food coloring

Package of dry yeast (found at the grocery store)

Small plastic cup

Measuring spoons

Funnel

Construction paper, markers, and some creativity

Safety glasses

Plastic tarp to cover the demonstration table

Rubber gloves

LET'S TRY IT!

1. Let's start with the arts and crafts part of the activity by making a decorative wrap to cover the plastic soda bottle. Since the activity is called Exploding Toothpaste, use your creativity to make a wrap that looks like a tube of toothpaste.

2. Put on your safety glasses and rubber gloves.

3. Cover the demonstration table with the plastic tarp.

4. Use a funnel to add 4 ounces (120 mL) of 40-volume hydrogen peroxide to the 1-liter soda bottle.

5. Add a squirt of dish soap and some food coloring to the hydrogen peroxide in the bottle. Give the solution a quick swirl to mix the contents.

6. Carefully cover the bottle with the toothpaste wrap that you made previously. It's best to have someone help you with this step to prevent you from accidentally tipping over the bottle.

7. The next step is to prepare a kid-friendly catalyst for the reaction by mixing an entire package of dry yeast with 4 tablespoons of very warm water in a small plastic cup. Stir the mixture with a spoon. If the mixture is too thick or paste-like, add a small amount of warm water to thin it out.

8. Here comes the fun part. Pour the yeast mixture into the bottle and watch what happens. It may take a few seconds to react, but the result is well worth the wait.

When you are finished, it is safe to dispose of all of the demonstration materials either by throwing them away in the trash can or by washing them down the drain.

WHAT'S GOING ON HERE?

Similar to what happened in the adult version of Exploding Toothpaste, the yeast works as a catalyst to release the oxygen molecules from the hydrogen peroxide solution. The oxygen-filled bubbles, which make up the foam, are actually the remainder of what happens when the hydrogen peroxide breaks down into water (H_2O) and oxygen (O_2). The bottle will feel warm to the touch because this is an **exothermic** reaction in which energy, in the form of heat, is given off.

BEYOND THE FIZZ: HOW TO GET KIDS EXCITED ABOUT DOING REAL SCIENCE

No one cared that it was cold outside. These kids could hardly wait to see what would happen next. Giggles and laughter bounced from child to child as the group of second graders positioned themselves around the 2-liter bottle of diet soda.

In a whispered voice, one boy asked, "Do you really think she's going to do it?"

"Sure . . . she'll do it, but you have to get ready to run," replied the girl standing next to him.

Mrs. Schmidt removed the roll of Mentos from her pocket and loaded them into a small tube attached to the top of the soda bottle. The only thing that kept the mints from falling into the soda was a plastic pin tied to a piece of string.

"Are you ready?" Mrs. Schmidt asked.

"YES!" shouted the students who could hardly contain themselves.

"Three . . . Two . . . One . . . Go!"

It all happened in a fraction of a second. Mrs. Schmidt pulled the string, the Mentos fell into the soda, and a giant soda geyser shot up everywhere. It was raining Diet Coke! As soon as the soda started to spray, the children scattered.

The students screamed, "That was awesome . . . do it again!"

When Mrs. Schmidt finally regained control, she told the children, "This is just a 'one-time' experiment. I don't have any more soda, but wasn't that cool?"

As she walked her students back into the classroom, Mrs. Schmidt knew that she had hit a home run with her exploding soda experiment. It had all of the elements of a great science lesson: it was engaging and the fun factor was huge.

Okay, But Where's the Science?

Like many teachers, Mrs. Schmidt thought that she had presented a great science lesson with her soda geyser activity. In her mind, she was doing an exciting, hands-on activity and her students were having a blast. However, if you look at the activity on a deeper learning level, you'll start to uncover the most important missing element . . . the act of actually doing science! The harsh reality is that Mrs. Schmidt's flying soda activity was cool, but her students were never doing science. The only level of student engagement was running away from the flying soda. The bottom line is that the students watched their teacher perform a cool trick using mints and soda, but calling that science is wrong on many levels.

Missing Elements . . . Wonder, Discovery, and Exploration

The first key to engaging students in doing real science is to understand the difference between a *science demonstration* and a *hands-on science experiment*. Demonstrations are usually performed by the teacher and typically illustrate a science concept. Science experiments, on the other hand, give participants the opportunity to pose *their own* "what if . . ." questions, which inevitably lead to controlling a variable—that is, changing some aspect of the procedure or the materials used to perform the experiment.

In Mrs. Schmidt's case, the students were never given the opportunity to ask questions, make changes, create their hypotheses, or compare the results of the new experiment with the original. When the students yelled, "Do it again," this should have been music to Mrs. Schmidt's ears. The great Mentos Geyser experiment captured her students' interest, and they were, in essence, begging for an opportunity to explore, to ask their own questions, to test changes to the procedure, to formulate new ideas, and to make their own big discoveries.

Instead, Mrs. Schmidt gave a commonly used response when her students wanted to be engaged: "No, this is a 'one-time' experiment." One time? Who can eat just one potato chip? No one ever performs an experiment just once! Demonstrations may be one-time events, but great experiments lead to more questions, which lead to making changes and trying the experiment again. It's a wonderful cycle of critical thinking called scientific inquiry, and you don't need a PhD in rocket science to pull it off.

Great Demonstrations Lead to Greater Questions

One of the attributes of an amazing science teacher is to watch how he or she uses a cool science demonstration to grab the students' attention and stimulate their natural curiosity. Great science teachers use demonstrations in such a way that they invariably precipitate the question, "How did you do that?" Evidence shows that students retain science concepts much longer when they observe an engaging demonstration that provokes an inquisitive response and that challenges them to figure out why. If the science demonstration served its intended purpose, the students will come alive with a stream of questions, and it's the job of a great teacher to help the young scientists turn their questions into an unforgettable learning experience.

> GREAT SCIENCE TEACHERS USE DEMONSTRATIONS TO INVARIABLY PRECIPITATE THE QUESTION, **"HOW DID YOU DO THAT?"**

Beyond the Fizz

No one can ever fault Mrs. Schmidt for sharing the Mentos Geyser with her students. If her primary goal was to get her students excited about science, she did it. But I hope she will discover a much more valuable treasure when she gives her students the opportunity to engage in the learning process. In the hands of a great teacher, cool science demonstrations like the Mentos Geyser open the door to an amazing journey filled with wonder, discovery, and exploration. By using the power of inquiry to create unforgettable learning experiences, you rekindle a childlike sense of wonder—in both your students and yourself—right before you turn them into a soaking mess from flying soda. Don't worry . . . they'll be talking about it at the dinner table for years to come.

WHO CAME UP **WITH THIS STUFF?**

Chances are, you've seen Steve Spangler . . . *because his picture is on the cover of this book!* Prior to flipping through the pages of this book, however, you might have watched one of his viral science experiments on television, surfing online, or observing your kids turn the kitchen into a mad scientist's laboratory. His incredibly popular Mentos-soda geyser experiment literally propelled his Colorado-based company, SteveSpanglerScience.com, into a new orbit as the go-to place for cool science toys and educational resources for making learning fun.

> As a hidden bonus for buying this book (and reading the back page), enter the special offer code FIREBUBBLES when you make your next purchase at SteveSpanglerScience.com.

What's it like to work at the Spangler labs in Englewood, Colorado? Anyone who works with Steve is quick to tell you that there's nothing ordinary about his or her job. When Steve Spangler is in the building, things are likely to fizz, pop, smoke, explode, or catch on fire . . . and that's just at the morning meeting. It's not uncommon for employees to walk on broken glass, touch 50,000 volts of electricity, or help test out a new experiment as Steve prepares for an upcoming appearance on *The Ellen DeGeneres Show*, where his contagious zeal for science reaches perhaps its widest audience.

Hands-on Science Workshops for Teachers

Steve Spangler started his career as an elementary science teacher in Centennial, Colorado, back in the early 1990s, and he continues to keep his classroom connection today as a professional development coach for teachers throughout the country. You can learn more about his training seminars and hands-on science workshops for teachers by visiting **SpanglerSeminars.com**.

Free Science Experiments and Cool Videos

If you've enjoyed the activities in this book, you'll probably spend more than just a few hours online when you visit **SteveSpanglerScience.com**. Since 2003, this website has been one of the leading sources for science fair project ideas, classroom experiments for teachers, and the home of one of the largest collections of science experiment videos online today.

Get a Behind-the-Scenes Look on Facebook, Twitter, and YouTube

Steve Spangler's unique teaching strategies and attention-getting creations have earned thousands of loyal Twitter and Facebook fans and millions of views each month on the Steve Spangler Science YouTube channel. Pick your favorite method of social media and follow Steve to see what he's got up his sleeve.